THE PRICE OF COLLAPSE

The Price of Collapse

THE LITTLE ICE AGE AND
THE FALL OF MING CHINA

TIMOTHY BROOK

PRINCETON UNIVERSITY PRESS

PRINCETON & OXFORD

Published by Princeton University Press
41 William Street, Princeton, New Jersey 08540
99 Banbury Road, Oxford OX2 6JX

press.princeton.edu

All Rights Reserved

ISBN 978-0-691-25040-3
ISBN (e-book) 978-0-691-25370-1

British Library Cataloging-in-Publication Data is available

Editorial: Priya Nelson and Emma Wagh
Production Editorial: Jill Harris
Jacket Design: Haley Chung
Production: Danielle Amatucci
Publicity: Alyssa Sanford and Carmen Jimenez
Copyeditor: Kathleen Kageff

Jacket image: Dai Jin, *Returning home through the snow*, 1455. Purchase, John M. Crawford Jr. Bequest, 1992 / Metropolitan Museum of Art

This book has been composed in Arno

Printed on acid-free paper. ∞

Printed in the United States of America

10 9 8 7 6 5 4 3 2 1

CONTENTS

TABLES AND FIGURES

Tables

vii

Figures

My Brief Life as a Price Historian

WE LIVE IN A WORLD that feels as though it is in the grip of rapid and capricious change. To rescue ourselves from the distress and dismay that change can induce, we tell ourselves that flux is the signature of contemporary life and sets us apart from the simpler worlds in which those before us lived. The speed and scale of climate change, price inflation, and political avarice over the past decade are producing greater turmoil than those of us who have lived long lives have ever experienced. Yet we really have little ground to be so confident that present flux is outdoing past, for there have been times when the very conditions of survival were stripped from our predecessors, denying them the dignity of living well. This book is about one of those times, China in the early 1640s, when massive climate cooling, pandemic, and military invasion sent millions to their deaths.

What happened in the early 1640s was a phase of the long global period of lower temperatures known as the Little Ice Age. Climate historians working from European data first dated the onset of the Little Ice Age to the 1580s, when Europe, like China, was plunged suddenly into colder weather. There is now broad agreement that this cold period began in the fourteenth century.[1] Late in the 1630s, the Little Ice Age began to plunge toward an even colder phase, inaugurating what is called the Maunder Minimum in honor of astronomers Annie and Walter Maunder, who hypothesized a link between the earth's temperatures and a decrease in sunspot activity, which they dated to 1645–1715.

This downturn precipitated the collapse of the Ming Great State (1368–1644), to give the Zhu family's dynasty its formal title. The Ming had survived with reasonable stability and durability for close to three centuries of the Little Ice Age.[2] It was not singlehandedly destroyed by climate, but its collapse cannot be explained in the absence of climate and human responses to climate. The dynasty responded, but its responses were overwhelmed during the deep downturn in the 1630s and 1640s. The story I tell here is not the tale of political disorder and military conflict leading to the suicide of the last Ming emperor and the Manchu invasion in 1644, a story that is well known to students of the Ming dynasty.[3] I offer the dynasty's fall instead as the closing moment in a two-century-long sequence of subsistence crises that pushed the people of the Ming toward a chaos that they could explain to themselves only as Heaven's scourge. In telling this story differently, I shall largely set aside the political events, factional feuds, and armed incursions across borders that usually dominate the narrative of Ming history to focus instead on data so ordinary that we take them for granted, prices.

I am not a price historian by training, nor a climate historian, for that matter. If I have been drawn to these fields in the latter phase of my career of analyzing historical change in China since the thirteenth century, it is because my concerns shifted to understanding China in a context greater than itself, which was not much in fashion when I started my studies. As it happens, the fields of price history and climate history were only just taking off when I entered graduate school in the 1970s.[4] The inauguration of Chinese price history could be credited to one of my graduate school mentors at Harvard University, Yang Lien-sheng, though as it happens, prices were never a subject of our conversation or study. When Professor Yang published *Money and Credit in China* in 1952, disarmingly subtitled *A Short History*, he gave us the first sustained work in English on money and credit. The book approached money not in relation to numismatics, as previous scholars had done, but in terms of its role in the economy and public finance. His conclusion that "the limited development of money and credit reflects the nature of traditional China" now sounds too sweeping in its cultural claims and unnecessarily pessimistic in its assessment of Chinese financial capacity, but it was a beginning.[5] Because he developed his findings regarding financial institutions in China in the light of Europe's experience, Professor Yang conceded a certain theoretical priority to the European record of money and credit. Still, his conviction that a common methodology could place both China and Europe within the same frame was a welcome push-back against the old tendency, in China as much as in the West, to regard China as an exception to what happened elsewhere. If his book offers little on the question of prices, it is because Professor Yang was interested more in money as a medium of account than in the real exchanges that money made possible. He was concerned to examine what money was in an institutional sense, not what it bought.

While Professor Yang was writing his history for English-speaking readers, Peng Xinwei, a Shanghai banker nine years his senior, was working on his magnum opus *Zhongguo huobi shi* (A monetary history of China). Peng's exhaustive study, published two years later, remains the basic handbook for any student entering the field of Chinese monetary history and continues to be cited, mostly in its revised 1958 edition as well as in Edward Kaplan's 1994 English translation. Instead of Yang's focus on public finance, Peng relied on numismatic studies to get to the question of prices. He approached prices as data for reconstructing and testing the changing value of money rather than as indicators of how people experienced what we call the economy.[6] From Peng's perspective, prices depended on the value of money and had no independent value on their own. My concern, by contrast, is to understand what prices meant to the calculations and strategies of the people who had to pay them. Monetary and price history are different but complementary exercises designed to produce different insights.

My introduction to price history, though I didn't realize it at the time, should have been at the University of Tokyo, where I went in 1979 to pursue doctoral research. Among the pleasures of those two years was the friendship of a young research scholar, Nakayama Mio as she then was before taking her married name of Kishimoto. The year we met, Nakayama published two superb essays on seventeenth-century price history, one a study in Japanese of commodity prices in the Lower Yangzi region, the other a study in English of grain prices in the same region and era.[7] Distracted by my own research topics, I failed to read these important works at the time and had no inkling that I would follow in her footsteps many decades later.

Prices did not really catch my attention until the 1990s when Denis Twitchett invited me to contribute an essay on Ming commerce to *The Cambridge History of China*. I intended to include price information in that essay, but I found too little material to work with. Even so, the 1990s was a good decade to start thinking about prices as my cohort of China historians began to situate our research in the comparative and connective contexts of world history. In 1998 the theorist of underdevelopment Andre Gunder Frank set a Chinese cat among the European pigeons with his spirited polemic *ReOrient* challenging historians to leave behind the Eurocentrism of existing models and think from the perspective of Asia. Gunder's attendance as an auditor in my graduate seminar at the University of Toronto made that challenge all the more personal and immediate. Central to his argument was the role that silver played in linking regional trade systems into a global network of commodity exchange moderated through prices.[8] Spaniards oversaw the mining of American silver, Chinese produced textiles and porcelains of unsurpassed quality and low price, and traders from all over got into the business of facilitating the exchange. The model was refreshingly ambitious and persuasively simple, and though it has since been criticized for simplifying a more complex network of relationships, it was a call to arms for those of us who wanted to situate China within a more global, more connected history.[9]

In this context, a quartet of books appeared at the end of the millennium to engage in that reorientation: Richard von Glahn's *Fountain of Fortune*, R. Bin Wong's *China Transformed*, my *The Confusions of Pleasure*, and Kenneth Pomeranz's *The Great Divergence*. Published between 1996 and 2000, these four books helped to bring China into global history free of the old Eurocentrisms. We did not examine prices closely, but we did ask price-related questions. Would a knowledge of prices enable us to compare the economies of China and Europe? Might Chinese price data help determine the degree to which China's economy influenced prices in the global economy? What role did these prices have in enabling Japanese and Europeans to enter trade networks? Or to put this question more bluntly, did Ming prices mean that Ming merchants were laundering the loot of Spanish conquistadors and Japanese warlords when they exchanged manufactures for the metal? We had no answers, but at least we had questions.

My focus on the culture of consumption and social investment in Ming China was what led me to prices. Once I began to find them, I looked up from the texts I was reading and realized that the history of consumption was pointing me not just to price history but to climate history as well, for it was in periods of climate disturbance that prices rose and chroniclers thought to write them down. This book presents a synthesis of what I have found. It is not so much a history of Ming prices as an account of the role that prices played in mediating the relationship between the people of the Ming and the climate that turned against them. While most of the documents were penned and published by the Ming elite, my goal has been to catch sight of ordinary people so as to better understand the decisions they made as they bought and sold their goods and services, especially in periods when China slid from prosperity to calamity.

Over the years spent researching Ming prices, I have accumulated a debt of thanks to students and colleagues who supported the work. For sharing the tedium of prying prices out of Ming sources, I am pleased to thank my former students Dale Bender, Desmond Cheung, Lianbin Dai, Si-yen Fei, Yongling Lu, Tim Sedo, Frederik Vermote, and Niping Yan. For sending me data or helping me think through issues in this book, I thank my colleagues Gregory Blue, Cynthia Brokaw, Jerry Brotton, Peter and Rosemary Grant, Robert Hegel, Geoffrey Parker, Bruce Rusk, Richard Unger, Pierre-Étienne Will, and Bin Wong. I reserve special thanks for Richard von Glahn for his careful reading of the final manuscript.

Financial support at an early stage of the project came from the John Simon Guggenheim Memorial Foundation. Some of the material in this book was presented as the Edwin O. Reischauer Lectures at Harvard University in 2010, followed later that year by a second set of lectures at the Collège de France through the kind invitation of Pierre-Étienne Will. I am grateful as well to Burkhard Schnepel who in 2016 invited me to pursue research on global trade prices at the Max Planck Institute for Social Anthropology in Halle, and to Dagmar Schäfer who three years later hosted me as a visiting scholar at the Max Planck Institute for the History of Science in Berlin where, with the help of Shih-Pei Chen and Calvin Yeh and access through the LoGaRT (Local Gazetteers Research Tools) project to sources made available via Staatsbibliothek zu Berlin's CrossAsia portal, I carried out certain technical analyses that would otherwise have been beyond my reach.

I wish to acknowledge with much gratitude three editors who have influenced my writing career: Sophie Bajard, who suggested that I write a book on Ming environmental history even if this is not quite what she had in mind; Kathleen McDermott, whose frank response to an earlier draft saved me from publishing a book that no one would want to read and put me on course to write the book I actually wanted to write; and Priya Nelson, whose enthusiastic support for my work has given me a new publishing home at Princeton University Press. Finally, if the book reads well, it is because Fay Sims had the interest and patience to listen to me read the final manuscript aloud and stop me when the prose stumbled.

THE PRICE OF COLLAPSE

1

The Tale of Chen Qide

> I was born too late to have witnessed the glories of the founding emperors Hongwu and Yongle [1368–1424] or to have seen the brilliance of the eras of Emperors Chenghua and Hongzhi [1465–1505]. (8a)[1]

Chen Qide (pronounced Chun Chee-duh) late in life was always looking back with dismay. Not having lived during the reigns of famous emperors in the distant Ming past counts among his lesser regrets, but this is the sentiment with which he opens the essay that he signed on the Ghost Festival—the fifteenth day of the seventh lunar month of the fourteenth year of the reign of the Chongzhen emperor, or 21 August 1641 by our calendar. It was the first of two essays he would write that year and the following to describe the collapse of his world.

The world he watched collapse was his home county on the Yangzi delta, Tongxiang, a hundred kilometers southwest of Shanghai. Chen was a local schoolteacher of no noteworthy accomplishment and would be utterly forgotten today were it not for the two essays in which he describes living through the Ming dynasty's worst years of disaster. That this memoir should even have survived is remarkable. It was not put into print until 1813, when a local historian appended it to a slim volume of Chen's advice on living an ethical life titled *Chuixun puyu* (Simple words handed down to instruct). To my knowledge only one copy of *Simple Words* exists today, in the library of the University of Nanjing. That his two essays are known to historians is because the editor of the 1877 edition of the Tongxiang county gazetteer, a locally published chronicle of county affairs, appended them to his chronology of local events at the end of the Ming. By such slender threads has Chen's account of the early 1640s survived across four centuries to feature in this book—and shape its tone and content. The text that follows intersperses passages from Chen's writing with my own commentary.

Having opened his memoir with this rather conventional piety about the greatness of earlier generations of Ming emperors, Chen Qide shifts from abstract regret to personal nostalgia.

> My memory goes back only to the opening years of the Wanli era [1573–1620] when I was just a boy and people always enjoyed bountiful harvests and prosperity. (8a)

Of Chen Qide himself we know only what he reveals in his writings, such as in this passage, which points to his having been born around 1570. He tells us something of his family life in a short piece in *Simple Words* entitled "The Ten Foundations of my Happy Life." There he delights in his good fortune of having been born into the topmost of the traditional "four categories of the people" (gentry, farmers, artisans, and merchants, in descending order of status) and of having been raised in "a family that plowed and read," meaning an educated rural family of middling wealth, which historians of the Ming call the local gentry. These privileges encouraged him to believe that "the world was an expansive place without restrictions."[2] Like other gentry youth, Chen studied hard in hope of grasping the glory of passing the civil service examinations and serving the state. Despite sitting the exam every three years through his twenties, he never managed to pass. At the age of thirty he abandoned that ambition and decided to get on with his life, turning to a career in teaching, which he pursued contentedly for the next twenty-five years. Chen was thus a responsible member of a modest, educated family comfortably nested in the lower ranks of the county elite.

To prove that the Wanli era was a time of abundance, Chen offers his best evidence:

> The price of a peck of grain never rose above 3 or 4 cents. (8a)

The term Chen uses for grain is the generic word *mi*, meaning kernel or seed. On the Yangzi delta, *mi* generally meant rice, the grain southerners preferred over the millet and wheat of northerners. He prices rice in a unit called a *dou*, literally a scoop or bucket. Converted to Western equivalents, a *dou* is 1.2 pecks, 2.8 US gallons, or 10.7 liters. Peck is not a precise equivalent, but I use it to translate *dou* in this book. The currency Chen cites is the standard small unit of silver, *fen*, meaning "one one-hundredth," in this case, one one-hundredth of a weight known as a tael, which is the dominant accounting unit of Chinese currency.[3] The Chinese word is *liang*, but we call it tael thanks to Portuguese borrowing in the sixteenth century of the Malay word *tahil*, meaning simply "weight." A tael of silver weighed 1.3 ounces or 37.3 grams, roughly the weight of a pencil. To translate *fen*, I have opted for "cent," the literal meaning of which is one one-hundredth, in our case, of a dollar. This translation is nonstandard, and so I must remind readers that a cent of Ming silver, though only a third of a gram, was not a trivial amount. When Chen was a child, it was enough to buy a gallon of rice (for an overview of weights and currencies, please refer to table 1.1 in appendix A, "Units of Measurement").

When prices were that low, prosperity was within the reach of all, or so Chen liked to think. No one needed to worry about not having enough to eat.

> Anyone who wanted to ferment grain to make liquor just threw away the dregs. Beans and wheat were fit only to feed oxen and swine. As for fresh fish and the choicest meat, every household had all it needed. People supposed that this prosperity would go on forever. (8a)

Knowing what was coming, Chen then shades his image of easy living with a warning that prosperity can produce not only well-being but moral laxity.

> How was anyone to know that people's hearts would give way to debauchery or that Heaven would deplore the surfeit? In the blink of an eye, the year *wuzi* [1588] arrived. Drenching rains soaked us, turning near and far into one vast slough. In the following year *sichou* [1589], the earth was parched for hundreds of miles. For two months, there was not a spoonful of water in the rivers, nothing but rank weeds. (8a)

Chen Qide's memory is exactly correct. The two years 1588–89 were a time of massive natural disasters, first flooding rains, then severe drought. Being a Confucian moralist, the only way Chen could explain this double wave of disasters was to find the human failing that caused it. In his view, the plenty of the early Wanli years had frozen the people's moral compass. Rain and drought were not simply natural disasters: they were Heaven's warnings. To quantify the severity of this warning, Chen returns to the price of rice.

> At that time, if you could put together a picul [10 pecks] of grain, you could get a price of 1.6 taels of silver. The price of grain soared and stayed high for months on end. Not a blade of grass was to be seen in the fields nor a strip of bark on the trees. Refugees filled the roads, and corpses sprawled in the streets. (8a)

Pushed to a price of 16 cents per peck, rice was four to five times what it had been. At that price, the poor turned to whatever substitutes they could strip from the natural world around them, from grass to tree bark. Social disorder followed.

After the disaster of 1589, Chen jumps forward to the 1620s, when the rise of the eunuch faction under the directionless Emperor Tianqi (r. 1621–27) threw the Ming regime into a crisis of leadership, to the deep dismay of political elites as well as moralists. Chen briefly describes the disarray of that period, presenting it as a warning from Heaven, then turns to what all along had been his destination, the late years of the reign of Emperor Chongzhen (r. 1628–44) (for the titles and dates of the reigns of Ming emperors, please refer to table 1.2 in appendix B, "Reign Eras of the Ming Dynasty, 1368–1644").

> When the thirteenth year of the Chongzhen era [1640] arrived, heavy rains fell for months on end. Floodwaters rose at least two feet higher than in the *wuzi* year of the Wanli era [1588]. The entire landscape in all directions became a great swamp. Boatmen poled their way among beds and couches while fish and shrimps swam through wells and stoves. Those with upper stories resorted to them as bolt-holes while those without scrambled onto their roofs or climbed up onto

terraces, having no thought in the morning but whether they would survive till the evening. (8b)

Again, Chen turns to the price of rice to quantify the disaster and track its course.

The price of grain started at over 1 tael per hectoliter [10 pecks] and gradually rose to over 2 taels. After the floodwaters receded, farmers from Wuxing [the neighboring county to the northwest] fanned out across the fields of Jiaxing [the prefecture of which Tongxiang was one county] searching for sprouts, which they fought over as though these were delicacies. Not until the end of the seventh month [early September] did they depart in an unbroken line of barges. (8b)

Here Chen moves up the decimal register from peck to *shi*, literally "stone," which is loosely equivalent to a hectoliter. Dividing by ten to derive a peck price, the price in 1640 rose first to 10 cents per peck, then to 20 cents.

As happened in 1588–89, flood gave way to drought the following year. The drought of 1641 was so severe that the riverbeds ran dry and prices were pushed up to yet another level.

The price of a hectoliter of grain was driven from 2 taels up to 3. Rural people had to pay 40 cents of silver for only 1 peck. Even though the spring wheat harvest was double what it had been in earlier years, in the end it was still not enough to feed everyone. Some ate chaff, some chewed bran, some even savored weeds and tree bark as though these were dinner and dined on chaff as an appetizer. (8b)

The economic effect of these prices was to shut down the markets. The social effects were disastrous.

A respectable family that could feed every member two meals of flour gruel a day celebrated their great good fortune, though the vast majority had to get by on one meal a day. Husbands abandoned their wives, and fathers their sons, each fleeing in different directions in the hope of surviving. Useful objects accumulated in the markets but went unsold, as those who could have used them simply walked away after asking what they cost. As for works of art and fine curios, no one even stopped to ask how much. Oh, the people's poverty was extreme. (8b)

Even the pawnshops were shuttered because no one had anything left to pawn. Farmers still robust enough to work went out to plant crops, though no sooner had they done so than locusts blanketed the fields and ate everything that sprouted. The streams ran dry, and there was no water to scoop into buckets and pour on the fields. Then an epidemic broke out, likely the plague, passing through the population and infecting 50–60 percent of households. A wave of suicide—Chen calls it "going to a tree"—followed.

Chen then steers his narrative back to prices, moving from the price of rice to the price of other foods to press his point about how prices kept going up to ever more impossible levels.

Not a single everyday item was less than several times more expensive than it had been before. Laying hens and geese were worth four to five times more than before. Even soybeans could not be bought for less than several dozen coppers. (9a)

The only food too cheap to be priced in silver was soybeans, grown to make tofu. Chen prices them in the thin minted bronze coins with a hole in the middle that were used to make small purchases. They were called *wen*, meaning "script," a reference to the reign title of the emperor during whose reign the coin was minted, which appeared on the obverse side of the coin. Again following Malay usage, Portuguese traders called this coin *caixa* (or *caxa* in Spanish), from which comes the English word "cash."[4] As cash has another meaning in English, in this book I use the old English term for a bronze coin of low value, "copper," to translate *wen*. The value of a thousand coppers was declared the nominal equivalent to one tael of silver at the start of the Ming, though copper immediately appreciated against silver, resulting in an exchange rate that hovered around seven hundred coppers per tael. Seven coppers were thus equal to one silver cent. One copper was not much money: it bought a block of tofu, a sheet of ordinary writing paper, two pairs of chopsticks, or a pound of charcoal.[5] Two coins purchased a cheap writing brush, a stick of cypress-wood incense, or a rice-flour cake. The poorest watched their coppers closely, whereas the rich would not stoop to pick one up off the ground—unless it was to make a point about trifling profits as distinct from real values.[6] To care about one copper coin became a contemptuous figure of speech for those obsessed with money.[7] A self-respecting Buddhist mendicant would refuse one coin as alms, accepting nothing less than several dozen.[8] The fact that it took several dozen coins to buy a single block of tofu in 1641 attests that a single copper on its own was close to useless during the famine.

At these prices everything that lived was eaten.

> As a result, a family of eight, unable to feed themselves, treasured pig fodder. Middling families could not afford to raise one pig, birthing sows having long been sold off to pay expenses. Previously a stewing pig might fetch a full silver tael, but now just the pig's head cost eight- or nine-tenths of a tael. As a result, whereas in the past you heard the constant noise of chickens and dogs, now you heard them only in the markets, and then only if you listened carefully. (9a)

Chen concludes by warning readers not to treat his memoir as "the confused ramblings of an old man." They should be grateful for having survived terrible times but should not forget the loss and suffering, as though nothing terrible had happened, which was what most people wanted to do.

The disaster was not over. When Chen put down his pen on the Ghost Festival of 1641, he could not have guessed that matters would go from bad to worse. A year plus a month later, on the Mid-Autumn Festival, or 19 September 1642, Chen put pen to paper and picked up the story where he had left off the previous year. He opens with the severe shortage of rice as winter came on. Rather than high prices, there were no prices because there was no rice to which to attach a price.

> At this time there was no rice in the market to buy. Even if a dealer had grain, people passed by without asking the price. The rich were reduced to scrounging for beans or wheat, the poor for chaff or rotting garbage. Being able to buy a few pecks of chaff or bark was ecstasy. Come the spring of the fifteenth year of Chongzhen [1642], the

countryside swarmed with people digging up the first green shoots as they sprouted. Earlier they had been selective about which plants they ate; by this point there wasn't a plant they wouldn't eat. Rural people filled their baskets and brought them in on their carrying poles, and in an instant what they brought was gone. Never had vegetables sold as fast as this. (9b)

The starving abandoned or butchered their children. The epidemic infection rate rose to 90 percent. Desperate to halt the downward spiral, people sacrificed what little food they could scrounge to the spirits in the hope of divine intervention, especially after the epidemic returned. This regrettable practice, Chen notes, only drove prices higher.

Because the infected made rich offerings [of food to the gods], food was twice as expensive as it had been the year before. A large chicken including both drumsticks fetched 1,000 coppers. A young cock barely able to crow was still worth 500 to 600 coppers. Stewing pigs went for 5 taels, then 6, even 7. A piglet was worth from 1½ to 1⅗ taels, even as much as 1⅘ taels. By contrast, a serving girl was worth only 1,000 to 2,000 coppers. Shouldn't people be valuable and domestic livestock cheap? (9b)

Here Chen cites prices in both currencies, chickens in copper, pigs in silver, and then humans in copper. Behind the use of coppers versus silver lay a widely understood social distinction, which was that copper was for cheap things and silver for more important purchases. By pricing a serving girl in coppers, Chen is implying that chickens should be priced in coppers but not people. In this collapsed economy, prices put pigs above people.

Not until the end of the summer of 1642 when the rice crop was harvested did the disaster start to ease.

The price of grain gradually went down. The sick started to recover, and the people regained their color, but they were saddened that those who died could not come back to life and that not all who had fled would return. (10a)

Chen concludes his second essay as he did his first, urging readers to take what happened as a warning from Heaven and pleading with them not to forget what had happened. It had been Heaven's punishment, and Heaven could punish again. As he concludes,

> Alas, in an evil time like this when misfortune strikes in successive years, not to die of starvation or disease must be counted as boundless good fortune. But if we fail to heed these warnings or to show gratitude to Heaven and Earth above and to our ancestors below, and instead just congratulate ourselves that we have survived the famine and then turn our attention entirely to acquisition and enjoyment, how can we recover our humanity? Unable to let what happened be forgotten, I have written this, my second account. (10a)

Finding a moral lesson in misfortune is a common response to hard times, especially when the commentator is a moralist such as Chen, who constantly patrolled the boundaries of the slight privilege he enjoyed at the lower edge of the county gentry, lest he slip from what he regarded as that bit of good fortune. His parents may have primed him for this mission by giving him the name Qide, which can loosely be translated as "a person of his virtue." With limited wealth and no examination credentials, Chen had only his virtue to maintain his position in society, and he was vigilant about putting it to use. As he writes in *Simple Words*, "Making yourself one one-hundredth smarter"—and here he uses the term "one *fen*," as though intelligence could be counted out just like hundredths of a tael of silver—"is not as good as reducing your attachment to everyday affairs by one one-hundredth." Or to phrase the same point differently, "Increasing your ability by one level is not as good as ridding yourself of one level of stupidity." This leads to his conclusion: "To be absorbed in worldly matters betrays a vulgar character; to tolerate your own stupidity is low class."[9]

Confucianism so closely linked ethics with cosmology that the boundary between them was almost nonexistent. Rain came from Heaven; if rain did not fall, it was because Heaven chose not to send it as a warning or a punishment. We live within a different cosmology, though even we take disruptions in the weather and the ecology of disease as morally charged warnings of environmental degradation and climate change. So we do not stand at all that great a distance from the people of the Ming, even though our moral calculations are very differently based. My intention in this book is to recover the world of the people of the Ming by meeting them halfway. We conceptualize the world as a physical ecosystem that is vulnerable to the effects of changing conditions, they as a metaphysical board game in which Heaven directs the action. The architecture of ideas is not the same, and I see no reason to adopt Confucian logic, but I do see reason in getting as close as we can to the experience and understanding of those who experienced that time and understood it in terms meaningful to them. Without attending to what a subsistence crisis meant to them, we hollow out the past.

In fact, we and they inhabit a global ecosystem that was and is prone to disturbance, whether because human folly blocks Heaven's blessing or human-generated carbon and aerosols block the sun's energy. We also share the habit of tracking our fortunes through the prices we have to pay. This book follows Chen Qide's lead and tracks grain prices, not as a barometer of Heaven's displeasure but as a measure of climate change. Consider it an extended footnote to Chen's account of the disasters of 1640–42.

A Ming Understanding of Prices

It might help to launch this footnote by asking what Chen and his contemporaries understood when they talked about prices. Believing that the world was best when nothing changed, they hoped prices would do the same. Everyone knew that prices moved seasonally in response to supply and demand, and that in normal times that movement should bring a price back to where it had earlier been, not push it up to a new, higher level. The language that expressed this aspiration for stability was that a price should be "level" (*ping*) or stable. Stability was important not just for anticipating how to manage the cost of living, but because an unstable price was unfair to someone. Price unfairness offended against the quality they called *gongping*, "level for all," which is to say in the public interest of all (*gong*) and treating all equally (*ping*).[10] A fair price was one that both buyer and seller could accept because it met a shared expectation. An abnormal price favoring one side over the other was not fair because the exchange created a relationship in which one person benefited at the expense of the other.

The best of all possible worlds was one in which "grain should be plentiful and prices fair, so that the crowing of roosters and the barking of dogs echo each other" between one prosperous and self-sufficient village and the next. According to this model, "people should not face the trouble of having to ship things over great distances in order to garner adequate profits."[11] These are the words of Shanghai native Xu Guangqi. Xu was one of a small number of educated Chinese who converted to Catholicism early in the seventeenth century (he took the baptismal name of Paolo). His knowledge of European Christianity equipped him to draw on both Christian and Confucian commitments to serve his emperor, rising eventually to the highest post of chief grand secretary just before his death in 1633. Paolo Xu wrote his comment about an economy that did not have to rely on long-distance trade in the mid-1620s at a time when court politics were in disarray but climate-driven calamities had sufficiently paused that people could begin again to imagine how the world would be best ordered. His intention was not to deny the necessity of commercial exchange, as some Confucian fundamentalists might have done, but to imagine a system in which commercial exchange brought benefits in which everyone could share, and which kept prices fair.

Merchants of the time offered the same argument—that trade leveled prices—to justify their work. As a declaration posted in a town square just west of Shanghai to protect the cloth merchants' guild put it in 1638, members of the guild "swapped prices to make a living." They bought at one price and sold at another, "always buying at a fair price and selling at a fair price, not cheating even a child," so that people could enjoy access to goods not immediately available to them. They wished to be seen as virtuous men performing an essential economic function, not as parasitic exploiters of producers or consumers. As the public declaration phrased this, they "set their prices according to the quality of the goods so that people can exchange what they have for what they lack."[12] Public sentiment was not always persuaded, especially that year as China edged toward the worst climate collapse of the millennium, but the argument nonetheless enjoyed some credit.

Chinese were not alone in believing that prices should be stable and that trade could be a means of achieving this fairness. Europeans developed an analogous discourse of just prices based on the conviction that prices should be stable, while also recognizing that of all prices, grain was usually the most volatile.[13] The idea that keeping prices stable and fair was not just a positive benefit but an act of justice was a view popular among seventeenth-century English advocates of trade. Rather than favor "the inequality whereby the one thing exceeds the other," pricing should "endeavor to bring them both to an equality," as Gerard de Malynes wrote in his handbook on commercial practices published in 1622, in the same decade that Paolo Xu offered his vision of an agricultural economy that commerce serviced in the common interest. "Equality is nothing else but a mutual voluntary estimation of things made in good order and true." Reasoning in similar terms, draper William Scott in 1635 declared prices to be the true measure of things in a just system. "As time is the measure of business, so is price of wares. If the price exceed the worth of the thing, or the thing exceed the price, the equality of justice is taken away." Economic historian Craig Muldrew explains that these writers believed that "a fair and just price was also the cheapest price." Fair prices served to "ensure that a free supply of goods was available equally to all, so that goods could be afforded by the poor as well as the rich." For it to be fair, a price had to be cheap enough for everyone to afford, yet also "high enough so that profits could be made, without which no one could have afforded to purchase anything else."[14]

The people of the Ming understood the logic of letting supply and demand determine fair prices, though they would have thought of this mechanism more in terms of the exercise of personal virtue than the working of abstract justice. Granting agency to the market as an autonomous generator of justice would not have been an argument that Ming observers were quite willing to make. The Confucian allergy to the use of the word "profit" to do anything but condemn private benefit over public good induced Ming writers to veer away from the idea that profit could be a mechanism for achieving justice. Prices were fair to the extent that both sides of the exchange benefited and neither profited at the expense of the other. In that context, people allowed that an open market was conducive to fairness. Accordingly, when speaking of a price that was fair, the term they often used was "market price" or "current price." To record that a transaction was carried out on the basis of market prices was to register approval of that transaction.[15]

Fairness did not rule out the reasonable use of market forces to ensure that the producer was fully recompensed for his products. The farmer with crops to sell who "waited out the time in order to wait for the price" was not condemned for doing so, simply understood as acting reasonably in his own interest within the terms of market exchange to obtain what for him was a fair price for his harvest.[16] So too when a merchant moved goods from markets where they were cheap to markets where they were more expensive, so long as he conformed to prevailing current prices in both markets, he was understood as serving the public good even as he garnered a percentage for himself. "Regardless of whether we are talking about state grain or people's grain, the price should always rise and fall with the current price and should not be forced up or down," argued famine-relief expert Yu Sen. That way, "when the price is high, merchants coming from far away will naturally be numerous. Grain being abundant, the price will level itself." This was how private commerce could serve the public interest. Yu accounts for its capacity to do so not by invoking the magic of the market, as his English counterparts might, but by referencing the Confucian notion that extremes tended to move toward the middle. As he concludes, simply, this movement "is something that the innate tendency of things makes inexorably so."[17]

The Presence of the State

Confucian thinkers such as Chen Qide, clinging to the lower edge of his county's social elite, concerned themselves with unfair prices and the limits of people's ability to afford them in times of crisis. Not always confident that markets would deliver goods at affordable prices, however, they held the view that when prices became unfair, the government should intervene to *ping* them, to "level" them to the amounts that people expected and could afford to pay. This is what the Ming state did, through several different mechanisms.

The most basic mechanism to ensure the affordability of grain was reporting. Local magistrates were tasked with monitoring grain prices in their counties, sending agents into the local markets every ten days to record prices and look for signs that they might be rising. Magistrates then forwarded this information monthly to the capital so that the court could keep abreast of food supply conditions across the realm.[18] In Beijing, the underlings assigned this task were agents of the dreaded Eastern Depot, the Imperial Household's intelligence agency. These men descended on markets in the capital on the last day of every month to check the prices of rice and other grains, and also of beans and cooking oil. The data they gathered allowed the court to know whether agriculture was flourishing. They could also be examined to determine whether merchants might be blocking commercial circulation of grain in order to drive up prices.[19] Though merchants were eager to represent themselves as fair dealers, no one quite trusted them in an economy in which there was little price transparency beyond what the next dealer down the street was charging. Coupled with this distrust was the inevitable relaxation over time of state scrutiny of prices. Institutional entropy led in 1552 to reducing the monthly price check to twice a year, in February and August.[20] By the 1570s, this mechanism seems to have fallen out of practice in most places except Beijing, where market price surveillance continued at least into the 1630s in order to anticipate threats to the stability of the capital.

In a crisis, the Ming state could intervene in ways more aggressive than data collection.[21] Sometimes it mandated the prices at which commodities should be sold. For example, an imperial edict issued during a famine in 1444 imposed an official price below the market price and required merchants to sell their grain at this reduced price.[22] When famine struck the Beijing region two decades later, however, the court adopted a different approach. Rather than setting a price, the capital censor issued a strong warning against price gouging. "In towns in many areas there may be strongmen and brokers who monopolize businesses and markets and make the price of grain and other commodities expensive when they should be cheap, and cheap when they should be expensive, entirely in order to make big profits and enrich themselves,"[23] he warned. Local officials should take action when they discovered that prices were being manipulated in such ways. An edict of 1523 repeated this warning more gently, reminding shopkeepers that "those who demand unfair prices" will be punished.[24] At the local level, judges had a free hand to prosecute merchants engaging in fraudulent pricing.[25] It was accepted in principle that merchants were the agents of healthy markets, yet as one magistrate wryly observed, "letting a broker in the market set commodity prices" would be like "allowing students to monitor their own misdeeds in school."[26]

More often than setting grain prices, the state intervened to influence them by releasing government grain stocks onto the market at prices that would compel grain merchants to lower the market price. Officials could also intervene by imposing blockades to prevent dealers from taking grain out of distressed economies to locations where they could make an even greater profit. "By imposing a ban at the right time," an official in South Zhili (the metropolitan province around the southern capital of Nanjing) advised a subordinate in the 1540s, "the price of rice will remain fair and the people have more than enough food for their use."[27] A century later, during the crisis that Chen Qide has described, a grand coordinator in South Zhili (where Chen's county was located) issued a proclamation imposing this ban in Suzhou. "The rice grown here under ordinary circumstances is not sufficient to meet local demand," he stated in his public proclamation. "Not only do parts of Zhejiang rely on shipping rice in from Jiangxi and Huguang provinces, but Suzhou also looks to this rice as the manna on which it survives." The grand coordinator reminded grain merchants of an earlier provincial prohibition "forbidding dealers from elsewhere from taking grain stored in the city of Suzhou and selling it elsewhere at a high price, leaving Suzhou empty as a result." He also ordered merchants further up the Yangzi River to keep the supply of grain flowing.[28] The history of Ming prices thus cannot be written in the absence of Ming state actors.[29]

The Ming state mattered to the prevailing price regime not only because it could intervene to affect prices, but also because it was a major buyer in the economy. The early Ming state to some extent bypassed prices by relying on requisitions and corvée (forced labor) to meet its needs for goods and services. Still, the government needed to make purchases, and the founding emperor was adamant that his officials pay for these purchases at market rates so as not to drive dealers into bankruptcy or interfere with the prices ordinary people paid. He even tried in 1397 to impose a law that made selling goods in the capital above "current value" a capital crime. His son, Emperor Yongle, reiterated this measure during his first year on the throne in 1403, extending the rule not just to the capital but everywhere throughout the realm right down to the villages.[30] These and other laws found their way into the Ming Code, the imperial law book, which scheduled punishments in relation to how much the price a seller charged deviated from the fair market price.[31]

By the sixteenth century, the state's involvement in the market was vastly greater as a result of shifting from requisitions and corvée to purchasing goods and services on the market with silver raised through taxation, and always, at least in principle, at market rates.[32] Underlying all these rules was the fundamental Confucian commitment to the idea that agents of the state should not act to compromise or erode the welfare of the people. In practice, of course, price exchange was a rough-and-tumble zone where officials, like everyone else, were out for themselves. For Confucian moralists this was a constant worry, just as it was for their Christian contemporaries. In his study of money and prices from that period, historian Jacques Le Goff notes the moralists' anxiety about the loss of true value in the face of money. If anything served to constrain the reduction of everything to a price, it was the Christian virtue of caritas, the care of others that was part of what was termed the care of souls, including one's own, so as to earn sufficient virtue to be admitted to Heaven. As long as caritas was in play, the abstraction of money was not powerful enough to clear a way for capitalism to override the obligations of justice.[33] People of the Ming would have recognized the virtue of such constraint, though they would have understood it in the Confucian terms of mutual obligation rather than the salvation of souls. The notion that money was part of the arrangements by which human beings were subordinated to the grace of God, however, they would have found puzzling, even incoherent. Where the people of the Ming shared a certain anxiety with their European contemporaries, it was over the danger of allowing greed to overwhelm obligation and letting money and prices to drive out reciprocity and alienate the poor from the rich. "The lord of silver rules Heaven and the god of copper cash reigns over the earth," as one disgruntled Confucian magistrate railed in 1609. "Avarice is without limit, flesh injures bone, everything is for personal pleasure, and nothing can be let slip," he continued. "In dealings with others, everything is recompensed down to the last hair."[34] The moral limits that should constrain economic behavior were entirely absent when dealers set prices.

At both ends of Eurasia, then, people experiencing the early phase of the economic growth that marked what is now called early modernity worried that price calculations and the pursuit of wealth were eclipsing calculations of virtue and care. As Chen Qide warned his readers at the close of his second essay, "If we just congratulate ourselves that we have survived the famine and then turn our attentions entirely to acquisition and enjoyment, how can we recover our humanity?" Wild prices were a warning.

Prices as Data

The task of this book is to scale up Chen's two years of price history in Tongxiang county to the national and dynastic level. It won't be easy. Chen had the advantage of living within the price regime that he knew as intimately and in as complete detail as we know the prices of everything around us in our own price regime. He could simply mention a price and expect his readers to know how far it diverged from what was fair and what that divergence meant. We therefore have to make up for our ignorance of his world by reconstructing the price regime within which he lived. Determining what things cost in the Ming period sounds like a simple technical task and a modest intellectual goal. It is neither.

Price records in a precapitalist economy are difficult to find, haphazard when found, and always incomplete.

Despite these difficulties, one thing that makes the reconstruction of the price history of Ming China possible is the attention that people of the time paid to prices. They knew that they inhabited a world in which everything—beans and rice, hens and maidservants, careers and survival—carried a price tag. When a palace eunuch offered to buy a military hero's sword in 1570, the soldier responded by demanding to know how such a thing "could be treated as a commodity," and yet the price offered was too attractive. He jettisoned his rhetoric and let the sale go ahead.[35] While moralists might insist that there were things that were not for sale and could not be priced, most regarded this view as quaintly sentimental at best, self-deluding at worst.

People kept themselves apprised of what things cost so that they would know what to buy or sell, and when and where to do so, or sometimes just to preserve a record of what they spent. A few even entered these prices in their diaries, letters, and reports, which is the haystack where the historian can go to find these needles. An instructive example is a text to commemorate the hanging in 1612 of a new bronze bell at Miyin Monastery in Chen Qide's county of Tongxiang. The text has survived because the author, a retired official named Li Le from the city of Hangzhou some sixty kilometers to the southwest, included it in his commonplace book *Jianwen zaji* (Notes on things I have seen and heard).[36] Li explains that he and others mounted the project to replace the original bell that the provincial government had confiscated and melted down to make firearms in 1544 during one of the waves of piracy in those years. He includes many figures in this text: the minimum gift that would get a donor's name into the official register of donations (3 taels); the amount of silver he and his friends raised in the first two months (200 taels) and then in the third month (another 200 taels); the amount of silver one of them took to the Ministry of Works in Nanjing for clearance to purchase over 1,600 kilograms of copper and tin (270 taels), which also yielded a ministry call to merchants to make the metal available "at a fair price"; a ministry authorization to waive tolls when shipping the metal to Tongxiang (which Li writes was worth "at least 60 taels");

the cost of restoring the brickwork on the bell tower (16 taels); reimbursement for costs borne by the monks at Miyin (30 taels); the bell caster's fee (35 taels); the cost of buying and inscribing the stone stele on which this text was incised (10 taels); and finally, an offering to Wenchang, the god of literate culture, to ensure the success of the project (40 taels). Li does not list enough figures to put together a complete account, though that was not his purpose. Posting the sums in full public view was a bid to protect the investment against later encroachment or theft.

While the exact reporting of actual costs and prices that this sort of purpose encouraged has left a vast archive of Ming prices, the data are scattered, inconsistent, and not easily summarized into statistics. This fact rather works against the initial appeal of prices as apparently hard facts and reliable data. When economic historian Earl Hamilton asserted in 1944 that prices are "the oldest continuous objective economic data in existence," he did so on the basis of the considerable surviving records of prices in Europe.[37] Hamilton was confident that the prices he found in European documents could be used not just to track price changes but to rewrite the narrative of historical change in the early modern world. To an extent, his confidence proved to be well placed in terms of yielding insights into historical change, yet no set of price data is without its ambiguities. Some prices may meet Hamilton's standard, yet the reader will see that prices are rarely the sturdy facts they appear to be.

Take the price of a Ming bucket, for instance. A Beijing magistrate recorded that in 1577 his office bought buckets for three silver cents each.[38] Is that what a bucket cost in the Ming? Possibly, though not all buckets are the same. What is the price of one bucket in a certain place and time might not be the price of a bucket in another place and time. Then there is the price itself. Three cents may have been the price his office paid, but did it involve a surcharge or a discount? Was it actually paid in silver, or was it paid in copper coins, and if so, at what rate of exchange? I raise these questions not to undermine the viability of price history but simply to note that the concrete price that someone said they paid at a particular place and time may not be what the thing paid for actually cost. If we turn to the records of a magistrate in the Zhejiang hinterland fifteen years earlier, we find that his bucket price was four cents.[39] Did the price of buckets fall over those fifteen years? Unlikely. Were rural prices higher than urban? Again, unlikely. Did the Beijing office buy so many more buckets than the rural office that it was able to obtain a volume discount? Possibly so, since the Beijing office was obliged to supply many more agencies with buckets than was a magistrate's office in a rural town. Or was it, more simply, that the Beijing bucket was small, cheap, and poorly made compared to the Zhejiang bucket? We have no way of knowing. The most we can conclude is that a sixteenth-century Ming bucket cost between three and four cents.

If the price in silver of a thing is complicated by the multiplicity of things, so too is it troubled by the mutability of silver. As historian Bruce Rusk reminds us in his delightful essay on the culture of silver in Ming and Qing texts, using unminted bullion rather than sovereign coins complicated the experience of exchange hugely. Just like a bucket, silver did not exist solely in its abstract function as currency. The metal came in many degrees of purity, some acceptable and some not. It could be adulterated and faked by those who knew what they were doing. Whether buyer and seller both accepted a particular weight of silver as payment depended on the many noneconomic factors shaping the social nature of the exchange, most especially, trust and face. As Rusk writes, "Although the market of value of a lump of silver might seem to be the simple arithmetic product of its weight and its purity, these two inputs were not free-floating variables, even when they could be so treated for accounting purposes. Silver retained an inalienable physicality: to function as an abstraction, a piece of silver had to produce a consensus about its level of purity."[40] The physicality of silver certainly distorted (or at least influenced) prices, though we cannot possibly access the quality of silver used in any particular transaction four centuries ago. Every recorded price is thus potentially as troubled by the metal that changed hands in exchange for the good or service that was purchased as by the quality of that good or the nature of that service.

"Price history at its best," to quote Earl Hamilton again, is supposed to deal with "only prices and wages actually paid in an open market by agents free from political or ecclesiastical coercion."[41] While I might like to be Hamilton's "careful price historian" who "avoids distortions by differences in the quantity purchased, seasons of the year, conditions of sale, transportation costs, services rendered, hidden charges, supplementary wages in kind, and the like," the documentary record does not allow me to meet that standard. Most Ming price data reach us not from the "open market" but from administrative records kept at the local level by government officials for the purpose of drawing up budgets and keeping accounts. They were supposed to follow market prices, though whether that happened the day they went to buy buckets, we cannot say. In the case of Ming China, most of the price data available to us were recorded at the point where state and economy met.

Forty years ago, historian Michel Cartier, recognizing the need for market prices to write price history, warned his colleagues about the perils of writing price history from such administrative documents. He doubted, in fact, that a history of Chinese prices was even possible before the eighteenth century.[42] Much of what survive in the Chinese record are in fact fiscal prices, that is, prices determined in the context of tax assessments. Fiscal price and market price may be in calling distance of each other, but not necessarily. Fiscal prices were often what Cartier termed "fossilized prices" set in relation to earlier market values and not subsequently adjusted. Sometimes these fossilized prices were replaced by artificial equivalents when local administrations converted their financial operations from taxes and salaries in kind to taxes and salaries in silver, as they did over the course of the sixteenth century, though how closely that conversion followed market prices is difficult to assess.

It is true that most of the prices I have recovered were compiled for noneconomic purposes: for the fiscal purpose of manipulating quota prices rather than to document market prices, or the administrative purpose of controlling graft, or the moral purpose of forcing prices into a fairer register, or the rhetorical purpose of narrating prosperity and decline. But price historians of China are not alone in facing these challenges. Five years before Hamilton made his cheerful pitch for the objectivity of price history, William Beveridge in the introduction to his history of prices in early modern England observed that "price history is a study not of isolated facts but of relations."[43] Prices are "the outcome of transactions" that arise in the context of a host of noneconomic factors.[44] In fact, it was often the relational character of prices rather than their purity as data that persuaded people in the past to write them down. This relationality is what makes them so valuable for the kind of history I offer in this book, which focuses more on understanding Ming society than Ming economy.

The Limits of the Possible

Two prices historians of Scotland have wisely cautioned that "prices and wages were real things that people paid and received, and on which the quality of their existence, and occasionally indeed their very lives, depended. The inner meaning and interpretation of the data may be subtle and difficult. But they are not merely numerical artifacts gathered to amuse and perplex historians."[45] The propensity of prices to amuse and perplex enticed not a few Ming authors to collect and tell outrageous price stories. Wondering at the shocking gaps between rich and poor, baffled by the rapidity of the changes overtaking their lives, they expressed their bafflement and outrage through price tales that anyone could understand.

If occasionally I follow their lead and indulge in a good story, I do so to highlight the fact that Ming writers deployed prices to mark not just what things cost, but how that cost did or did not make sense in relation to everything else they knew, and as well how prices shaped the social relationships that bound people together. Of all the price stories I can tell, the best is the one I have already told, Chen Qide's account of the turmoil in the last years of the Ming dynasty. It could be objected that choosing Chen's story of spectacular disaster as the centerpiece of the book's architecture drags us away from capturing everyday life. In fact, it sheds strong light on the core issue of survival, which is when prices really mattered. The coincidence of the Ming dynasty with the middle phase of the Little Ice Age only further encourages me to treat the disasters of the 1640s as the core around which to reimagine what it meant to live in the Ming world of prices. And what better price to focus on than the price of grain, the key of human survival.

Four decades ago, historian Fernand Braudel proposed that the first task in writing a complete history of the early modern world was to evaluate "the limits of what was possible."[46] He did this in the two opening chapters of *The Structures of Everyday Life: The Limits of the Possible* by addressing the two basic features of society that determine its capacity to manage those limits: population size and food supply. The balance between mouths to feed and grain to feed them can be a delicate one in an agrarian economy reliant principally on solar energy for food production. Working from European historical records gave Braudel certain advantages compared to those of us working on other parts of the world. His cohort of demographic, climate, and price historians had access to local parish and market records that allowed them to construct price series for many commodities. Availing themselves of such documents, it was possible for them to detect long-durational shifts and to model economic, social, and even political changes over the long run. Braudel's analysis would have been unthinkable without price history.

Whereas European historical prices are mostly raw numbers preserved in handwritten archives, Chinese prices are mostly printed outcomes that have reached us via the administrative filters through which they were processed. The difference means that we write different histories.[47] But this need not rule out the possibility of historical comparison. Historical China can and should be part of what we draw on to understand the Braudelian limits of the possible. Every premodern society worked within the physical constraints of growing enough food to feed the people, or to put that in more abstract terms, the capacity to capture and transform solar energy at a level that permitted a population to reproduce itself, even in good times to expand.[48] So long as people have left records of how they accomplished this task, and sometimes failed, price history is possible. As I shall show in chapter 4, the best documentary proxies in the Chinese record regarding changes in the energy output of the sun or the accumulation of aerosols in the earth's atmosphere to block that energy are the prices to which grain rose when that energy declined—which is the record to which Chen Qide made his small but moving contribution.

That the people of the Ming lived within a price regime does not mean that we can reduce their conviction that prices should be stable to the idea that the economy consists of fair exchanges between legal equals. As historian William Reddy has cautioned in the context of early modern Europe, prices in a commercial economy are neither conceptually abstract nor socially indifferent.[49] The idea that prices were innocent cogs in a system of just equivalences was a notion that British economic polemicists later in the seventeenth century would develop to disguise the asymmetry of monetary exchange that encouraged the growth of capitalism. A rich buyer and a poor buyer might pay the same price for something, but they parted with their money for that thing from different capacities.

Nascent economic theory in early modern Europe may have cast money as a disinterested medium by which legal equals traded goods and services, and prices as the objective means that fairly determined what should be paid, reducing prices to the mathematical abstractions by which buyers and sellers in a capitalist economy obtained a fair exchange. For those with limited means, however, prices were shackles at best, capricious furies at worst. They benefited the wealthy as they entered their sums into their account books, exploiting the neediness of the poor and exposing them to precarity. Prices became the means by which money created social structures, enabling the rich and disciplining the poor without either realizing or acknowledging how steeply tilted was the playing field of the economy in which they were positioned. A price may be "level" in relation to other prices, yet the field on which exchanges were performed through prices was anything but level. People of the Ming could agree that the best condition for achieving fair prices was an open market free of monopolistic dealers and conniving officials, but the ideology of capitalism—that capital was value-free and generative of universal benefits—did not follow; nor, for that matter, did the ideology of Confucianism as a system in which rich and poor provided mutual support. In the real world, prices arrayed rich against poor, and price-based exchanges continually reembedded unequal social relationships. Ensuring that prices were "fair" was not a bid to unshackle the economy from society. It was a bid to keep economy and society linked so that the financially advantaged and commercially astute did not deprive the poor of their means of survival, especially when climate crisis destroyed the crops in their fields.

Disaster Prices as Climate Proxies

What sets the greatest limits on what was possible in a preindustrial agrarian economy is climate. Crops require warmth and water to germinate and grow. When the supply of either is compromised, so is the production of food, and when production falls, prices rise. Grains are relatively hardy crops that can tolerate a certain measure of environmental stress, yet moisture and warmth have to stay within a certain range for the grain to sprout, grow, head, and mature. What determines temperatures and rainfall is how much energy the earth receives from the sun. The link between climate and grain prices in a preindustrial economy is fairly direct: from the amount of energy the earth receives from the sun, to the warmth and water that reach grain in the fields, to the price to buy that grain. Before the large-scale burning of hydrocarbons that has characterized industrialization, two factors could alter the short-term energy relationship between sun and earth. One is depressed solar radiation due to sunspots, or the sun's ejection of solar matter. The other, called climate forcing, is alterations in the atmosphere, whether from volcanic eruption or large-scale forest burning spewing aerosols into the atmosphere and blocking solar radiation from reaching the earth's surface. In both cases, though the mechanisms differ, the earth is deprived of its usual quantum of energy. This reduction in energy causes temperatures to fall and induces changes in the movements of winds and currents that alter patterns of precipitation.

Climate exerts an uneven impact on the earth's energy owing to regional variation in the distribution of land and water mass, yet the globe is shaped by the larger energy system that determines conditions on the earth. Climate is simultaneously local in its manifestations and global in its overall capacity and trends, though the one does not directly determine the other. The climate of a region cannot be automatically deduced from climate elsewhere. Nor can the pattern of the whole be fully constructed and understood without detailed knowledge of all local manifestations, which is why regional climate reconstruction for a zone as large as China is essential for refining and improving our knowledge of climate change.

Chinese scholars turned to the task of writing a history of China's climate in the 1930s by extracting references to natural disturbances in the dynastic histories, the official records produced after the fall of a regime by its successor. Since the 1990s, Chinese climate scientists have moved away from documentary proxies in preference for physical proxies that can be instrumentally measured. Their work has broadly confirmed that China's climate history shares global patterns derived from physical evidence collected elsewhere in the Northern Hemisphere—the Mediaeval Warm Period from the mid-tenth century to the mid-thirteenth, followed by the Little Ice Age starting in the fourteenth, deepening in the fifteenth, then sinking further through the late sixteenth to nineteenth. The most valuable physical proxies have been tree-ring data (the markings left by cellular growth indicating variations in temperature and precipitation) and glacial-ice-core analysis (of water molecules for their isotopic weight or the presence of volcanic sulfur). Both physical archives have been of great value for detecting to a high level of precision the amplitude of climate variation over time, as well as for detecting regional differences.[50] In recent decades, however, some climate scientists have begun working with historians to access climate information embedded in documentary materials, such as historical chronicles, diaries, and letters in which contemporaries recorded observations about unusual weather as they were experiencing it. These documentary proxies may lack the consistency of tree rings and water isotopes, but they tend to be consistent within a culture and can prove more sensitive—to local conditions, to moments of sudden change, and most valuably, to the impact that that change had on people's lives—than are tree rings.[51]

To this body of documentary proxies, I propose to add grain prices. As I began to detect a relationship between deviant grain prices and climate change in Ming China, I was surprised to discover that environmental history elsewhere has paid almost no attention to prices. A noteworthy exception is a short essay that economic historians Walter Bauernfeind and Ulrich Woitek published in 1999 analyzing grain prices in Germany during the Little Ice Age in relation to climate disturbances. They observed that "most recent research in economic history has rarely focused on the influence of climatic changes on important economic and demographic variables," and that historians have sought to explain the "price revolution" of the sixteenth century "by population growth, currency debasement, and growth of money supply," assigning climatic deterioration "only a minor role."[52] Even in *Global Crisis*, his magnum opus on the global climate disasters of the 1640s, Geoffrey Parker turned to proxies other than prices. When I asked him whether there might be price research buried in his footnotes that I had not noticed, he confirmed that indeed prices played no role in his masterwork.

The neglect of prices in historical research on climate change struck me when I began to notice records in Ming local gazetteers of the prices to which grain rose during climate disturbances. A correlation between climate fluctuation and price fluctuation became most persuasive to me as I found records in local gazetteers of famine grain prices in the 1450s. When I went looking for research that might corroborate the effect of climate change on prices in other parts of the world in this period, I was pleased to discover Bruce Campbell's references to studies of price movements in *The Great Transition*, in which he uses prices not so much to reconstruct climate shifts as to register the impact of environmental conditions.[53] Campbell pays particular attention on the decade of the 1450s, tagging those years as "the coldest decade of the fifteenth century and the coldest of the Middle Ages."[54] This was when the global decline in temperatures, known as the Spörer Minimum, which began in the 1420s, dragged the Little Ice Age down to its first extremely cold phase. That the 1450s was when famine price records first began to appear with some consistency in local Ming records emboldened me to consider their correlation with climate fluctuation at the global level. Further research through to the end of the dynasty confirmed the reasonability of working from this hypothesis that prices could be treated as climate proxies. Not being trained in econometrics, I have not mathematically modeled the correlation for Ming China as Bauernfeind and Woitek did for sixteenth-century Germany, but the strong coincidence between Ming famine prices and global climate disturbances gave me the confidence that readers will accept this connection as not just intuitively obvious but theoretically compelling.[55]

I shall return again and again in this book to Chen Qide's memoir of the price disaster of the early 1640s. To close this introductory chapter, I would like to comment on a curious detail that might help to bring us close to the world this book examines. Chen Qide's second essay is dated to the fifteenth day of the eighth lunar month, the Mid-Autumn Festival. One of the most important festivals of the Asian year, the Mid-Autumn Festival was when families and friends got together to celebrate the harvest. In the late Ming, farmers had a saying: "If the Autumn Equinox precedes the Mid-Autumn Festival, a peck of rice will trade for a peck of coppers. If the Autumn Equinox follows the Mid-Autumn Festival, a peck of rice will trade for a peck of beans."[56] The one price was extravagantly high, the other dismally low. The Mid-Autumn Festival is a lunar date, whereas the Autumn Equinox is a solar date, the day when the sun at noon stands directly above the equator before it starts to tilt away and push the Northern Hemisphere toward winter. What the saying did was enshroud the mystery of what price the harvest might bring within the mystery of how Heaven arranged the movements of the sun and moon. In fact, the equinox never precedes the full moon of the eighth month. It would be a miracle if it did, just as it would be a miracle if a farmer could sell his crop of rice for its weight in money instead of its weight in beans.

As it happens, when Chen signed his first essay in 1641, the two days were tantalizingly close to each other. That year, the equinox fell just three days after the full moon. In 1642, however, the Mid-Autumn Festival arrived one day earlier and the Autumn Equinox one day later, widening the gap by two days. Had a farmer in Tongxiang any rice to sell in 1641 or 1642, he could have sold it for its weight in copper coins, though only to the wealthy. For most people, it would have been a nominal price that no one could afford. China was sinking toward the Maunder Minimum. On this autumn equinox there was no harvest to celebrate and no crop to sell—which is now our task to understand.

2

Halcyon Days?
The Wanli Price Regime

CHEN QIDE remembered his childhood in the opening decades of the reign of Emperor Wanli as a time when foods was cheap and prices were fair. Wanli ascended the throne as an eager eight-year-old in 1572 and occupied it until he died, alienated and miserable, in 1620 at the age of fifty-seven, possibly from excessive use of the opium that his eunuch agents imported into Canton (Guangzhou) and shipped up to the palace in Beijing. Presiding over the Ming realm from youth to old age meant that he ruled through the many stages that a person goes through in a lifetime, which is why we find him responding differently to the challenges that ruling a population of one hundred million posed over the near half century he occupied the throne. There is therefore no single Emperor Wanli, just as his era cannot be reduced to a single static period. So much happened during his reign that the mere mention of the era produced a wide range of reactions, and still does. For some, it was a time of political factionalism, wasteful extravagance, and moral decline. For others, it was a period of social dynamism, philosophical renaissance, and economic prosperity. Though these characterizations veer to opposite sides of the road, both are true.

The reign of every emperor was given a title intended to capture its intention or spirit, and his was Wanli, which could be translated as Ten-Thousand-Year Agenda. Born Zhu Yijun, he was the legitimate child of his father, Emperor Longqing, but his mother was an imperial concubine, not empress, which normally would mean that he was not destined for the throne. When Longqing died in July 1572 at the young age of thirty-five, however, both of Zhu Yijun's elder half brothers, sons of the empress, were dead. The way was thus open for him to ascend the throne six weeks before his ninth birthday. As a teenager, Wanli was keen to learn his trade under the firm hand of a determined regent, Zhang Juzheng. After his mentor died in 1582, when the young emperor was in his twenties, Wanli took up the reins of government to rule the realm on his own.[1] Coddled within the walls of the Forbidden City, he suffered, as almost every emperor did, from a lack of exposure to the world his subjects inhabited. Still, as a young man, he seems to have done his best to stay informed about what was going on in the realm. Whether he knew what it cost to buy a bucket is another matter. But he was aware at least of famine prices. We know this from a conversation with Lady Zheng, his favorite consort four years his junior, on 19 April 1594, which he shared with his chief grand secretary the following day.[2]

Yesterday We were looking at *Jimin tushuo* [An album of the famished]. The Imperial Consort was in attendance.

"What are these pictures?" she wanted to know. "They show corpses, and people drowning themselves!"

"This is an album of pictures of famished people in Henan province, which Supervising Secretary Yang Dongming in the Ministry of Justice submitted to Us," We explained. "Right now that region is suffering disturbances due to a severe famine. Some people are eating tree bark, some devouring other people. He presented Us with this album so that We might know about this famine and swiftly organize the relief that will save those whose lives hang by a thread."[3]

This brief transcript is a remarkable record, not least for revealing what a conversation in the privacy of the palace might have sounded like. Whether the emperor and his consort used these exact words is less important than Wanli's assertion that they did. It was his way of making clear to his chief grand secretary and court officials that he was aware of the looming famine and would be personally involved in mounting the relief effort. Wanli would never have laid eyes on a famine victim, as soldiers were sent out to clear the streets he would travel of every distressful sight whenever he left the palace. He knew what famine looked like only from the album. He had received it some two weeks earlier, when a note in the court diary reports that the album had left him "shocked and distressed."[4] As it was still lying open on a table in his residence two weeks later, he must have scrutinized the pictures several times, shocked by what he saw, just as Lady Zheng was in her turn. By the time he showed her the pictures, he had had two weeks to read and digest the accompanying text that explained in some detail the sequence of human consequences that natural disaster could provoke, starting with social chaos and ending in mass violence.[5]

The album achieved exactly the outcome that the official who sent it into the palace hoped for: an immediate and full response. As Wanli told his chief grand secretary,

> When the Imperial Consort heard Us explain this, she chose of her own will to take what she had accumulated from funds presented to her over the years and make this available to rescue the people of this region.
>
> "I wonder whether this would be acceptable?" she asked Us.
>
> "Excellent!" We replied.
>
> The Imperial Consort gave 5,000 taels of silver in relief funds, but We thought that was too little and expressed the hope that We could count on her to make a further contribution, which she did.

Lady Zheng's gift was followed by grants from the empress dowager, two of the princes, and Wanli himself.[6] Their combined contributions amounted to the largest single famine relief effort that any Ming emperor undertook using Imperial Household funds. The family's generosity gave Wanli's chief grand secretary the moral leverage to require all officials above the fifth rank to donate their salaries to relieve the disaster.

What actually relieved the threat of famine was not these gifts, however. The solution was to encourage price differentials to do the work. When the price of a hectoliter of grain rose to the unheard-of price of five taels of silver, grain merchants were incentivized to transport grain down the Yellow River into the affected region. The grain barges were "lined up end to end on the incoming waterways for over forty *li*" (20 km), as the official in charge of the relief effort was able to report to the emperor.[7] As a result, the price of a hectoliter of rice fell to four-fifths of a tael. This was still high, but at roughly the level that the price might reach after a poor harvest, not in a major famine, and was sufficient to allow the merchants to more than cover their costs. As a result, none of the people facing starvation in Henan province died.[8]

The story is more complicated, in two ways. A close inspection of local records from Henan that year reveals that the province was not on the brink of famine. There was a mild downturn in grain production, but nothing like the scenes of devastation presented to the emperor. It turns out that the fear of famine was greater than its reality. If Wanli jumped the gun and made his officials jump with him, it was because of the memory of the terrible famine that had struck China six years earlier, which had caught the administration unprepared, and which Chen included in his history of local disasters.

The second complicating element in this story is Lady Zheng. Wanli favored her above the other women around him. After she became pregnant in 1586, he proposed to the Ministry of Rites that she be elevated to the status of Imperial Noble Consort and their son designated the heir apparent, even though he was third in line after Wanli's other two sons. The resulting succession dispute troubled Wanli for the rest of his reign, alienating him from his ministers and indeed from his job. By making Lady Zheng the hero of the story and then obliging his officials to line up to support her initiative, Wanli was doing his best to enhance her status and advance his campaign to have their son named his successor. The politics of this incident have nothing to do with grain prices, of course. It is simply to note that, at the Wanli court, politics always lurked just beyond the edges of every decision.

This chapter is not about either the person of Emperor Wanli or famine prices, which will be the subject of chapter 4. It is about the Ming world as it was lived during his reign, more particularly about the prices that constituted what I refer to as the Wanli price regime, and how people managed their lives within that regime. The question waiting in the wings is whether Chen Qide was justified in recalling the Wanli era as the halcyon age he wanted it to be.

Keeping Accounts

If 3 pecks of beans are bartered for 2 pecks of wheat, 10.5 pecks of wheat for 0.8 pecks of linseed, and 1.2 pecks of linseed for 1.8 pecks of rice, what is the price of each?[9]

This is a mathematical problem that a Ming adolescent with some education could solve. Once solved, however, the answer turns out to be ludicrous. Rice usually sold at a price slightly above wheat, sometimes as much as 15 to 20 percent higher. The solution to this problem, however, is that rice is 8¾ times the price of wheat, which was impossible. Now this is only a math problem and need have nothing to do with reality. I mention it simply to keep in view the fact that the people of the Ming conducted their affairs according to a reasonably high level of what historians of accounting call "economic utility."[10] They understood what things cost, how to compare prices, and what they had to do to make ends meet. They tracked "incoming" (*ru*) and "outgoing" (*chu*); they were keenly aware of what they "expended" (*yong*) and what they "retained" (*cun*). They disaggregated expenditures and income, understood that fixed assets had values, and could evaluate returns on investment. Ming price records would not exist were matters otherwise, especially in the Wanli era, the period of Ming history when price documentation palpably increases.

The increase in price records in the Wanli era reflects changes in the economy, which had become sufficiently commercialized for everything to have a price, and sufficiently productive that things from all over the realm and beyond were available—always for a price—in unprecedented volume and variety. Contemporaries marveled at the abundance that surrounded them. As the metaphysician and encyclopedist Song Yingxing declared in the preface of his survey of all things *Tiangong kaiwu* (The making of things by Heaven and humankind), he and his contemporaries "lived amid the world's ten thousand things and phenomena." He writes that "people know nothing about these things unless each of them, one by one, is placed before their eyes or explained to them,"[11] which was his purpose in compiling his book. Li Rihua, a Wanli-era art collector, doubted that the number ten thousand in fact did justice to the full extent of what existed. He believed that "in the space between Heaven and earth, novel things emerge with time. There is no original number of them that we can determine."[12] Things already known were best understood by reading what the ancients wrote about them, whereas new things, which were constantly reaching Li from places as distant as Venice and Lima, had to be examined firsthand and then added to the treasure house of the world. And every one of those things had to be assigned an appropriate price that would settle where it sat within the universe of Ming things.

The prices of basic commodities in the Wanli era were known to all. Animal protein appears in Ming records almost unvaryingly at 2 cents a catty. Jesuit priest Diego Pantoja confirms this standard price in a letter home in 1602, noting that "the best of all sorts of meat—beef, sheep, goslings, chicken, venison—cost 2 *liards* (a small French coin, which he uses to translate a cent of silver) for a *livre* (or pound, which is how he translates catty)."[13] Pantoja makes this observation because, by European standards, meat was in plentiful supply and prices, especially for game, were low.[14] Meat could go above 2 cents when quality was high, supply limited, or demand intense, yet these distortions were obvious to most people and of little mystery.

Price literacy became more challenging when the thing carrying the price was not an everyday food but a manufactured object. Consider armchairs. Whereas pork was always the same thing, more or less, there was no single Ming armchair, and hence no single price for a Ming armchair, only the many prices at which armchairs were made, bought, and sold at particular times and places.[15] As prices make sense only in relation to other prices, consumers relied on them to distinguish a fancy armchair from an ordinary one, as well as to distinguish themselves from buyers at other social levels.[16] Prices also created equivalences. People of the Ming could understand that a modestly priced armchair was price equivalent to a large crossbow or a turban fashioned from cut velvet, and that an expensive armchair was the same price as a porcelain dish of precious Ding ware or ten large geese.[17] To the extent that such price equivalences and differentials were not vulnerable to random alteration, a stable price system functioned. The burden on buyers was to know what things should cost, as well as what someone at their social level was expected to buy, and therefore the price at which they should buy it.

Returning now to Chen Qide's recollection that prices in the Wanli era were stable and universally beneficial, we need to test his conviction by reconstructing the Wanli price regime: what prices prevailed in this era, and how they varied in relation to each other. The data used in this chapter come from two bodies of sources. One is the corpus of essays, diaries, letters, and memoirs that proliferated during the Wanli era, texts in which authors sometimes refer, usually unsystematically, to prices as they encountered them in the course of their daily lives. The other, more systematic body of material, which I shall now introduce, are the official price lists and inventories that local magistrates compiled to guide, and limit, government spending.

Two County Magistrates

Appointment as a county magistrate was for most graduates of the state examination system the first rung on their career ladder in the civil service. Counties, of which there were roughly a thousand in the Ming, were the state's lowest, and arguably most important, administrative site. The county was where the state met the people, whether through legal hearings or tax collection or moral instruction or militia service. Taking young men who had been amply trained in the moral philosophy of the Confucian classics but not in the tasks of administration and dropping them into county posts where they were outsiders and did not speak the local dialect was trial by fire. The new magistrate had to learn quickly how to draw up and review budgets, raise revenue, and keep his office's financial affairs above water. The carrot for doing so was promotion; the stick was dismissal. Some flourished, many didn't, and most just listened to their local staff to balance the books.

Sorting out a county's financial records often meant having to reconstruct them from the ill-kept or nonexistent files that a careless or disaffected predecessor had left behind. That accomplished, the next task was to tally revenues and expenditures, on the books as well as off, to determine whether the one was in calling distance of the other, both to ensure that the magistrate could make ends meet and to satisfy the demands of reporting within the financial system.[18] The task after that was to lay down spending and revenue guidelines to keep the county afloat and the magistrate's own career on track. In this section, we shall meet two county magistrates who discovered financial chaos when they arrived in their posts, rose to the challenge, and left records of what they did—records laden with prices.

Hai Rui, born on Hainan Island in 1514, started his career two decades before Wanli took the throne. He became nationally prominent in state administration and politics through Wanli's first fifteen years as emperor. Shen Bang was born a quarter of a century after Hai, though he outlived him by only a decade. He did not rise to Hai's position, but he distinguished himself as a brilliantly competent county magistrate during the first quarter century of Wanli's rule. Intriguingly for what this says about the misalignment between the state curriculum and the examination system, the two had one thing in common. They were only provincial graduates, not holders of the highest title of *jinshi*, "presented scholar," which is to say, they were well-educated men who never made it to the national examinations. This was often the cohort where the Ministry of Personnel went looking for capable administrators. Hai and Shen were selected at the start of their careers for minor posts, and once they demonstrated their competence as administrators rather than as scholars, they both built impressive careers on the basis of their performance in office.

Our encounter with Hai Rui begins when he was serving as the magistrate of Chun'an. This impoverished town in Zhejiang, far upriver from the provincial capital of Hangzhou, was technically a subprefecture rather than a county, though administering such a unit was equivalent to running a county. When Hai Rui took up office in 1558, Chun'an had a reputation for lawlessness. It was a place where powerful locals had managed to intimidate or bribe the employees of the subprefectural office into getting their land holdings and fiscal obligations removed from the records. As soon as Hai was aware of that practice, he realized that the only way to run Chun'an without falling afoul of his superiors was either to press the common people for revenues beyond their legal obligation, which was what his predecessors had done, or to audit the books and correct the frauds that had pushed his office into desperate financial circumstances. Which path to take was never an issue for Hai. He saw his role as securing the interests of the people. "With every cent by which you reduce costs," he observed, "the people receive the gift of a cent." The outcome was a set of rules he called *Xingge tiaoli* (Regulations initiated and suspended), a title announcing that he was revising long-standing local practices by abolishing (*ge*) old precedents (*li*) one by one (*tiao*) and starting up (*xing*) new procedures.[19]

Hai prefaces his audit by writing that the first thing he did upon arriving in Chun'an was to go over the books. The exercise revealed that over half of the household registrations in the cadastral registers were "empty," which is to say that the household had fled or disappeared into other households, invariably with the connivance of corrupt clerks, leaving their tax obligations unpaid. The result was that the population of Chun'an on the books had fallen from over seventy-seven thousand in 1371 to forty-six thousand in 1552, a decline that clearly flew in the face of the reality of China's population growth in his own time.[20] The effect of their disappearance from the population registers was to shift the tax burden to those still on the registers. "In times like this, there is no way that change is possible," an old hand in his office warned Hai. "In times like this, nothing can be done. Better to go on oppressing the people than to excite the displeasure of your superiors. Better to strip those below you than to fail to indulge the great and the good."[21] Hai rejected this practical advice and launched the audit. Accounting, he believed, would reveal matters as they were, not as corrupt taxpayers or officials had altered them in their favor.

Another plank in Hai's reform was not to pay above market prices. He alerted local suppliers that they should not expect to get away with overcharging for goods and services, and he put his underlings on notice that they should not force suppliers to sell at below-market rates.[22] This intervention required posting prices, which Hai does in three different sections of his handbook: costs for holding state rituals in the "Rites" section, soldiers' food costs in the "War" section, and an inventory of the furnishings in the official residences of the magistrate (first class), the vice magistrate, registrar, and schoolmaster (second class), and the magistrate's chief assistant (third class).[23] Provincial rules required the local magistrate to furnish these residences, yet Chun'an had slipped into the bad habit of letting outgoing officials take the furnishings with them or pass them on to their subordinates rather than leave them behind.[24] Hai's inventory specifies how much should be paid to replace each item, should it need replacing, in an attempt to hold down costs. It provides a rare glimpse into the world of everyday prices for middling households (see table 2.1 in appendix C, "Tables for Reference").

Shen Bang faced a similar condition of financial chaos and shoddy accounting when he arrived to take up the magistracy of Beijing's Wanping county in 1590, although the specifics were very different because of the burdens that the revenue-hungry central government and palace placed on his operations. Just as Hai Rui did, Shen realized that an audit was the only way out of the crisis when he saw that his office was obliged to pay for costs of over 6,000 taels yet had a mere 52 taels in the treasury on his arrival. Shen had the right experience to manage this situation. Wanping was one of two counties into which the central government, following the logic of divide and rule, had split the capital. His previous posting had been Shangyuan county, which was one of the two counties into which Nanjing had been split by the same logic. Shen was used to managing the demands coming down from above. As his prefectural superior wrote in Shen's praise, his chief talent lay not in composing fine turns of phrase but in strategizing *jingji*, "the management of resources," the term we now use to translate "economy."[25]

The goal of Shen's audit was to enable him to create an annual county budget so that he could carry out the first principle of Ming accounting, which was to "measure revenue" in order to "determine expenditures." In fact, Shen had to look in both directions, keeping income up but doing everything he could to bring expenses down without falling afoul of the central government or the Imperial Household. "If there are no quotas for income and no controls on expenditures," he asked rhetorically, "how can anything be planned?"[26] To secure the findings of his audit, he decided to place all the figures in public view. The data appear in the two long chapters entitled "Expenses" in his *Wanshu zaji* (Unsystematic records from the Wanping county office), the administrative handbook for Wanping county he published in 1593.[27] The audit lists every item, from scallions to lacquerware, that the county had to buy to meet demands from agencies in the capital higher up the chain of command, including in most cases its price. Shen's handbook lists more prices than any other Ming publication. To quote an admiring superior, "the chapters on the county's expenses are the most detailed, to such a degree that the clerks have no way to profiteer." Some of the prices are what Shen terms "commuted prices," that is, silver payments based on the prices of real objects but made in lieu of supplying those objects. He also occasionally aggregates costs by listing a series of objects and assigning them a total cost. Even with such shortcuts, *Unsystematic Records* is unparalleled for the scale and quality of its data, providing page after page of prices for everything from cherries to chamber pots to chains for restraining criminals in court, and furnishing much of the price data that I rely on in this book.

For a Cent, a Mace, and a Tael

To summarize the Wanli price regime and make it tangible to us, I have drawn prices from the information that Hai Rui, Shen Bang, and others have recorded to assemble three tables. The first lists twenty-five things that could be bought for 1 cent of silver; the second, for 1 mace (or 10 cents) of silver; and the third, for a full tael (or 1.3 oz.) of silver.[28]

The first table displays items that could be purchased for a cent of silver (see table 2.2 in this chapter). This was a price that most people could afford. One cent was an almost universal price for a catty (1⅓ lb. or 600 grams) of vegetables, such as cucumbers, water chestnuts, and the like, though it was too much for basic cooking ingredients such as scallions and ginger, which sold for a fifth of a cent or less, which is to say, 1 or 2 coppers. One cent was enough for Hai Rui to buy a catty of

TABLE 2.2. Twenty-Five Things You Could Buy for a Fen (1 Cent) of Silver

		Place	Year	Source
Food and beverage (per catty unless otherwise noted)				
Cucumbers	黃瓜	Beijing	1590	SB 122
Walnuts	胡桃	Beijing	1590	SB 123
Liquor	酒	Chun'an	1560	HR 85, 86
Trout	鱒魚	Beijing	1602	DP 112
Eggs (18)	蛋	Beijing	1602	DP 112
Game		Huayin	1615	HB 305
Pork (half catty)	豬	Beijing	1590	SB 129, 130
Tobacco (half catty)	煙子	Beijing	1590	SB 134
Housewares				
Ladle	水瓢	Beijing	1577	SB 141?
Carrying pole	木杠	Beijing	1590	SB 133, 147
Stool	腳火凳	Chun'an	1560	HR 129, 132, 134
Fire tongs (pair)	火筯	Chun'an	1560	HR 130, 132
Whisk	茗箒	Beijing	1590	SB 151
Wicker basket	荊笆	Beijing	1590	SB 141
Porcelain soup bowl	礎湯碗	Beijing	1577	SB 141
Wok lid	鍋蓋	Beijing	1577	SB 141
Paper and ink				
Fruit-wrapping paper (25 sheets)	粘果紙	Beijing	1590	SB 171
2 exercise booklets	常考卷	Chun'an	1560	HR 42
Tablet of woodblock ink	寫版墨	Chun'an	1562	HR 83
Materials (per catty)				
Fish glue	魚膠	Jiangxi	1562	TS 164
Hemp rope	麻繩	Beijing	1590	SB 134
Tinfoil	錫箔	Beijing	1590	SB 133
Iron nails (half catty)	鐵釘	Beijing	1590	SB 145
Coal (10 catties)	煤	Beijing	1590	SB 140
Rosewood	紫檀	Hangzhou	1572	WSX 139

liquor, though liquor was available at multiple price levels. (Shen Bang provides prices ranging from 4 to 20 cents a bottle.)[29] As the standard price of meat was 2 cents per catty, 1 cent was only enough to buy half a catty (300 grams or two-thirds of a pound). Meat on the hoof, however, was half the price of butchered meat.[30] Fish, fresh or salted, usually sold at the same price as meat, though in a letter in 1602, Jesuit missionary Diego Pantoja wrote that you could buy a catty of trout in Beijing for 1 cent.[31] Tobacco, like meat, cost 1 cent for half a catty, according to Shen Bang. The Wanli era was when tobacco reached China from the Philippines, where the Spanish had introduced it, which is why it was still too expensive to buy a full catty at that price.[32] Shen's lists include several other tobacco prices, from 3 cents a catty for native tobacco to 10 cents for imported Philippine tobacco—which happen to be the earliest dated records of the consumption of tobacco in China.[33]

Minor housewares, from stools to wok lids, even cheap porcelain bowls, sold for one cent a piece. One cent was the lowest price at which you could buy ink and paper, again of the lowest quality. Hai Rui and Shen Bang both report prices for many types and qualities of paper, seventeen from the former and over sixty from the latter, just as one might expect in a government office that had constantly to generate reports and announcements, in units of one hundred sheets known in Chinese as a *dao* or "cut."[34] Finally, one cent was enough for a catty of such basic materials as glue, rope, and wood, though only enough for half a catty of iron nails.

Moving up from one cent to ten bought you foodstuffs at the high end of consumption: a catty of melons or giant peaches, 2 catties of yams, a whole large fish or fowl, and a catty of the finest imported tobacco and tea of the best quality (see table 2.3 in this chapter).[35] Like liquor, tea came in a wide range of qualities and prices, and these prices were stunningly unstable at the top end of the market where retailers competed for the business of tea connoisseurs.[36]

The housewares that you could buy for a mace are more elaborate than those that cost a cent, whether because of the costlier materials from which they were made, such as iron for a wok, or the work required to make them, such as a parasol. Paying ten cents moves you up to better-quality paper and ink, and to costlier materials such as candles, plaster, and sappanwood, an imported wood used to produce red dye. Items of clothing only begin to come onto the list at ten cents. So too medicine begins to appear at this level. Ten cents was also a common price for services, such as a doctor's visit. This price in indirectly

TABLE 2.3. Twenty-Five Things You Could Buy for a Mace (10 Cents) of Silver

		Place	Year	Source
Food and beverages (per catty unless otherwise noted)				
Winter melon	冬瓜	Beijing	1590	SB 122
Giant peaches	水蜜桃	Shanghai	1628	YMZ 170
Gorgon seeds	茨實	Beijing	1590	SB 123
Fine tea	細茶	Beijing	1590	SB 123, 126
Yams (2 catties)	山藥	Beijing	1590	SB 122
Carp (each)	鯽魚	Beijing	1590	SB 123
Hen (each)	雉雞	Beijing	1590	SB 122
Housewares				
Wok	鍋	Chun'an	1560	HR 130, 132, 134
Parasol	日傘	Chun'an	1560	HR 129,131
Toilet	淨桶	Beijing	1590	SB 147
Set of reins	套索	Beijing	1590	SB 132
Wooden bed	木床	Beijing	1577	SB 141
Four-poster bed	四柱床	Chun'an	1560	HR 129, 131
Paper (per 100 sheets) and ink				
Accordion-folded paper	抬連紙	Beijing	1590	SB 146
White report paper	白咨紙	Beijing	1590	SB 123, 129
375 g of liquid ink	墨	Chun'an	1562	HR 42
Materials				
Candles (per catty)	燭	Beijing	1572	SB 136, 137, 145
Sappanwood (per catty)	蘇木	Beijing	1590	SB 133
Plaster (per peck)	土粉	Beijing	1590	SB 133
Clothing and cloth goods				
Linen (per bolt)	苧麻	Beijing	1590	SB 138
Silk wrapper	段絹錦幅包袱	Jiangxi	1562	TS 161
Cloth bedcover	布被	Jiangxi	1562	TS 160
Embroidered kneesocks	繡護膝襪口	Jiangxi	1562	TS 161
Medicine (per tael)				
Aconite (wolf's bane)	附子	Shanghai	1620	YMZ 161
Services				
Doctor's visit		Shaoxing	1600	LW 131

confirmed by a reference in *Plum in a Golden Vase*, in which the novelist signals that a doctor, a character in the novel, was of mediocre reputation because he charges only 5 cents for a prescription. This point is underscored later in the novel when Ximen Qing, the novel's main character, pays a physician a full tael to save his favorite concubine—who happened to be the former wife of the 5-cent doctor.[37]

Moving up from a tenth to a full tael meant entering an entirely different level of consumption (see table 2.4 in this chapter). Little in the way of food and drink can be found at this level, as foodstuffs were too inexpensive to rise to that price. The only comestibles I have found for this table are a piglet and the highest-quality tea. Animals are the main exception. One tael was not quite enough to buy a whole animal. Fully grown pigs could cost up to 4 or 5 taels.[38] The furniture that a tael could buy was more elaborate and substantial, the fabric more luxurious, and the paper heavier and of higher quality than it was at ten cents. Poster paper (*bangzhi*), which came in large sheets a meter and a half long and almost the same in width, was the standard form of official paper in wide official use. Demand kept its price relatively high.

Several new categories of goods appear in table 2.4 that are not there in the two earlier tables. A tael bought you a firearm, and not just an ordinary fowling piece but an arquebus called Bilu, apparently Peru, suggesting that it was, or imitated, a Spanish type of gun. Another new category at this level is books. The Wanli era sustained an active print culture, with more readers buying more books, some to study for the exams, some to acquire knowledge for its own sake, and some simply for amusement. Prices varied depending on the quality of the paper and the wood used for the blocks, the fineness of the engraving, the skill with which the book was printed and assembled, and the accuracy of the editorial work. Books produced for popular consumption could sell for as little as 6 cents a volume, though these were thin fascicles rather than full titles.[39] Least expensive were the mass-produced and cheaply priced books that commercial publishers printed in Fujian. As bibliophile Hu Yinling summarized the range of book prices, "ten Fujian imprints are not as costly as seven from Zhejiang, which are not as costly as five from Suzhou, which are not as costly as three sold in Beijing." Of the

TABLE 2.4. Twenty-Five Things You Could Buy for a Tael (1.3 Oz.) of Silver

		Place	Year	Source
Food and beverage				
Piglet	豯豬	Beijing	1590	SB 122
Yixing tea (catty)	芥片茶	Shanghai	1620	YMZ 159
Furnishings				
Stove	爐	Shanghai	1592	PYD 307
Porcelain stove	宣爐	Shanghai	1593	PYD 308
Cinnabar-red chrysanthemum box	朱紅菊花果合	Huizhou	1518	WSX
Lacquered pearwood summer bed	素漆花梨木等涼牀	Jiangxi	1562	TS 160
Inkstone	硯	Shanghai	1591	PYD 307
Bonsai	盆景	Shanghai	1588	PYD 298
Set of 120 everyday porcelain dishes	磁器	Chun'an	1560	HR 131
Paper (100 sheets)				
Large poster paper	大榜紙	Beijing	1577	SB 139
Large porcelain-blue paper	大磁青紙	Beijing	1590	SB 145
Red pasting strips	貼紅籤紙	Chun'an	1560	HR 82
Letter paper	箋紙	Chun'an	1560	HR 82, 84
Textiles and cloth goods				
Tabby (bolt)	綢	Jiangxi	1562	TS 157
Silk bed curtains	錦段絹紗帳幔	Jiangxi	1562	TS 160
Silk coverlet	錦段綾絹被	Jiangxi	1562	TS 160
Materials				
Pearwood plank 3 m long	梨板	Beijing	1572	SB 138
Weapons				
Peruvian (Spanish-style) gun	密魯銃	Zhejiang	1601	LHL 6.66a
Books				
Primer of Tang poetry (4 vols)	李袁二先生精選唐詩訓解	Fujian	1618	SC 112
Epigraphic rhymes (6 vols)	廣金石韻府		1636	SC 112
Double encyclopedia (8 vols)	搜羅五車合併萬寶全書	Jianyang	1614	IS 263
Compendium of texts (20 vols)	新編事文類聚翰墨大全	Jianyang	1611	IS 263
Services				
Painting a set of new year's door gods	寫新歲門神	Nzhili	1548	WSQ 46
Mounting a painting	裱面	Shanghai	1591	PYD 299
People				
Boy singer	小廝	Shanghai	1588	PYD 289

seventy-eight book prices I have collected, the mean price works out to 2 taels. A quarter of the sample cost less than 1 tael. Forty percent were priced between 1 to 3 taels, a price band that can be taken to represent the middle range of fairly highbrow reading.[40] Above the mean, prices rise gradually to 10 taels, beyond which they leap to multiples of tens and hundreds of taels for collectors' items. We get a good glimpse of book prices in the Wanli-era library of Qi Chenghan. Qi told his sons told that the books he was leaving them as their inheritance were worth over 2,000 taels. The inventory he prepared of his collection lists 9,378 titles, which puts the average value per title at just under 20 cents.[41] What Qi's sons did with the books after he died is not known, though the attractive option for most people would have been to convert the books back into money, which is every collector's nightmare. A few decades later, Lu Wenheng, who struggled to rebuild his grandfather's Wanli-era library after it was lost to fire, was sufficiently cynical to anticipate that the next generation would be indifferent to such a legacy. "The sons who love wealth are many while the sons who love books are few," he lamented. "Collect them and hand them down, and if they don't fill the stomachs of bookworms, then they are good only for covering pickle jars."[42]

One tael is a level at which it became possible to purchase specialized labor, whether to commission a painting or have a painting mounted on a scroll. A tael could also purchase labor outright, enough at least to buy an adolescent boy or girl. One tael was generally the minimum. Prices ranged upward to 6 taels for a child, and from 4 to 20 taels for an adult, with sexual partners commanding a premium over domestic staff. Shanghai resident Pan Yunduan notes in his diary four prices he paid for domestic servants between 1590 and 1592 in a range from 4 to 10 taels.[43] Sometimes he bought couples. For Chen Wen and his wife, Pan paid 8 taels in 1592. For housepainter Gu Xiu and his wife, he paid only 2 taels in 1588, possibly because 1588 was a year of severe famine, when selling oneself into survival was better than starving to death. Pan also bought entertainers. Between 1588 and 1590 he bought nine male performers for prices ranging from 1 to 20 taels (an average price of 6.7 taels).

What do we learn from this exercise? Most importantly, that one tael was a lot of money for ordinary people, marking something of a dividing line between rich and poor.[44] A poor person might spend a tael on a piglet in hope of a future return, but he could afford nothing else in table 2.4. For every one of those items, the poor searched out cheaper substitutes. A pearwood bed for a tael was nothing the poor would seek to own. The rich might also not seek to own it, for a completely different reason. In the inventory of the wealthy clan that owned that particular bed, it was the cheapest bed they possessed. The most expensive cost five times that price.[45] From the perspective of the wealthy, a tael was not enough to spend on a bed.

Ming fiction often used one tael, in fact, as the boundary beyond which anything could happen. A tael was not quite enough to commit a serious crime, but anything over it was. In "Shen Xiu Causes Seven Deaths," a story Feng Menglong published in 1620, the final year of the Wanli era, a barrel cooper believes that he might be able to get 2 or even 3 taels for someone's pet thrush. He murders its owner and offers it to a merchant. The man's opening offer is only 1 tael, but the cooper haggles the price up by 20 percent and then rushes home to tell his wife the good news. They were both, writes Feng, "beside themselves with joy" at getting 1.2 taels.[46] Not by chance, the same boundary existed among religious donors. When a Buddhist monastery in Jiangxi was fundraising in the late 1610s, it set 2 taels as the minimum contribution for getting your name put on the public register of donors.[47] One tael was not enough to win the distinction of being publicly recognized as a patron of the Buddha.

A Spaniard in Canton

It is now time to situate the scattered data of the previous discussion in a lived context. For this purpose, I present two brief case studies to show how real households negotiated their way within the Ming price regime. The first case comes from urban Guangdong province, mostly Canton; the second, from a rural county at the southern end of the North China Plain. Both cases postdate the Wanli era by a few years, but they invoke prices that fall comfortably within the Wanli price regime.

The source for the Canton case is Adriano de las Cortes. The Spanish Jesuit missionary based in the Philippines washed ashore along the

Guangdong coast when on 16 February 1625, the Portuguese ship on which he was sailing to Macau was blown onto the rocks 350 kilometers short of its destination. Las Cortes was among the two hundred–plus sailors and passengers who were able to struggle ashore before dawn when the ship went down. Immediately apprehended by the local militia, they were conveyed, stage by stage, to the provincial capital, Canton. Las Cortes and his fellow hostages spent over a year in China before the provincial governor finally dismissed the charge of piracy against them and authorized their repatriation to Macau. Once back in Manila, Las Cortes wrote a detailed account of his experience, one of the finest ethnographies of Ming China written at the time, though it languished unread for centuries in an archive. Of outstanding reference value for us is chapter 20, "On the Wealth, Riches, and Poverty of the Chinese," in which Las Cortes describes the household economies of people he met.[48] He opens the chapter by cautioning the reader not to be misled by the glowing accounts of the prosperity of the Chinese economy that he had given earlier in his book. "The quantity of merchandise that the Chinese possess is not a sufficient argument to prove that they are very rich. In a general sense, this is, on the contrary, an extremely poor people." Las Cortes then proceeds to attend closely to the financial lives of the people he encountered in Guangdong, starting with hired laborers, and then soldiers, after whom he moves to shopkeepers and artisans, fishermen, and finally the elite. Describing the wealth of the Chinese "is difficult to do with exactitude," he warns, partly because of the severe poverty in which most people lived, and partly because of the invisibility of household wealth to outsiders. Nonetheless, he gives his Spanish readers an intimate sense of how Chinese lived, what they could afford, and how shockingly little they could get by on.

Las Cortes begins his account by observing the conditions under which ordinary people lived. Most people own "a small dog, a cat, a chicken, and a piglet."[49] Their diet consists of nothing more than "a little rice and some vegetables," except on festival days when even the extremely poor "do not fail to have a few mouthfuls of meat, fish, eggs, and wine in their style." Most of their food comes from the garden plots the men tend when they have not hired themselves out for work. As for the clothing of the household, it is augmented by no more than "two or three pieces of cloth a year." He sums up the total assets of an ordinary household in these terms: "If you add the little silver they might get for their clothing, the rest of their linen, and all their furniture, their total wealth would amount to barely 8 or 12 ducats," the Spanish unit of account that he uses to translate "tael." He concludes: "Among the mass of the poor, this represents a tidy sum, because many of them are even more destitute."

Las Cortes then turns to the livelihoods of the soldiers who were among his constant companions while he was under escort and house arrest, and whose lives he observed closely. Many were quite as poor as day laborers, though some, he writes, "have a bit more property, dressed and ate a little better, and were even able to be a bit more comfortably established and to have as much as 30 taels' worth of possessions." Like laborers, they fed themselves from their own plots. "The rest of their wealth consists of their salary, their furnishings, and other items." The captain's salary "is only four taels a month, without counting what each soldier gives him out of his pay for feeding the horse that he is obliged to have. On one occasion, I asked one of these captains the price of a good horse like the one he had, and he answered that it was worth 15 taels," a price that other horse prices confirm.[50] The monthly salary of ordinary soldiers, by contrast, was only "1 tael, plus a supplement according to their obligations," and that tael amounted to even less. Once the soldier had paid off his obligations, such as pitching in for his captain's fodder and paying off his other expenses, notably his gambling debts, "that tael reached his hands tithed and diminished."

Artisans and shopkeepers, Las Cortes observes, were better off than laborers and soldiers, though the financial range of this group was wide. "The capital of merchants who are well established and possess an ordinary shop selling various goods might with difficulty reach 30 taels, and of those with rich shops, 150. As for what artisans earn, one will discover that attributing to them a wealth of 30 taels is already too much."[51] To exemplify the household economy of a modest artisan, Las Cortes instances a tailor who was his neighbor in Canton. The man had previously done passably well as a soldier, then switched to tailoring. The priest knew him while he was running a successful business that attracted lucrative orders, though that business could be sustained only by his never taking a day off, even during festivals. "When I asked him what he could earn in a month, he replied, 1 tael. Out of that amount, he has to pay a yearly rent of 8 to 10 reals," which is how he translates "mace" (a tenth of a tael), "for his shop, and cover the needs of his family, his wife and children." The tailor's financial situation was not greatly different from a soldier's, yet to some extent he was better able to control his income and expenditures through hard work and close management. "As for other similar lines of work, I found out more or less the same thing; and for those who made a good earning, I saw that they make in a day 2 condimes," which is how he translates "cents." He stresses that "this is already quite a lot" in comparison with what an unskilled laborer is able to earn. "In that way an artisan can make about a quarter of a mace" a day. A daily take of 2½ cents is a modest income, but it gives "the advantage to the artisan of being able to add to the ordinary menu a small bit of salted fish, which as a further advantage is already luxurious."

Las Cortes marvels at how those in the lowest of trades were able to make a living, given that the products of their labor were so cheap. He offers the example of incense, which he found unimaginably cheap. No matter where he went, 10 cents bought between ten and twelve thousand sticks. Even allowing that Chinese incense was of poorer quality than Spanish, Las Cortes still judged the sticks to be of a respectable length and well made. When he tries to imagine how an incense maker can make a living at this price, he is baffled. "One has to suppose that the artisans who make these sticks earn something and are able to live from this industry, seeing as no one, or so it seems, can earn 1 mace from packaging 12,000 of these sticks."[52] The only conclusion he can draw is that Chinese have perfected the art of living at the barest level of subsistence, that they have learned how to make do with the tiniest of profits and to survive on "a bit of bad rice and a few leaves of mustard greens." He does not quite get to the proposition that poor households survived by pooling the marginal earnings of their members' labor, but he approaches that insight on the next page when he observes that "tiny gains" are how the majority of the poor get by.

The only other form of labor on which Las Cortes provides financial information is fishing. From what he saw, fishermen too lived at a bare subsistence. Among these men, he writes, "there are no fortunes of 30 taels, unless one counts a boat and nets."[53] The comment is instructive because it repeats the figure of 30 taels that Las Cortes keeps coming back to as something of a marker of the assets of better than the poorest household, a level that most fishing households could not meet. To support Las Cortes's account of fishing as a meager livelihood, I can offer a comment that a woman makes in a story collected by the Nanjing writer Zhang Yi. When her daughter-in-law turns to begging with a prancing monkey to earn some money, her impoverished mother-in-law advises her, "What you can make by begging with a monkey isn't much, and certainly not as good as fishing, which will earn you a daily profit several times more."[54] Fishing was no way to get rich, but it was better than begging.

A Magistrate on the North China Plain

Xun county sits at the southern end of the North China Plain, in what during the Ming was the southernmost toe of North Zhili where that province inserted itself between Shandong and Henan just north of the Yellow River. (During the Qing, Xun was reshuffled across the provincial border into Henan.) When the editor of the mid-Ming prefectural gazetteer characterized Xun people as both quick-witted yet trustworthy, he caught the measure of Xun's position at the edge of the commercial transformation that was altering the Ming. Xun people were still insulated from its worst effects (which meant that they were trustworthy) yet not entirely protected from its corrosion (which required them to be quick-witted).[55] While Las Cortes was in Guangdong recording prices in silver, financial transactions in Xun were recorded in both copper coins and units of silver, signaling that Xun was partly in and partly out of the commercial economy that tended to count prices in silver.

Xun county comes into view through the writings of Zhang Kentang, a native of Songjiang prefecture on the Yangzi delta southwest of Shanghai. Zhang was posted to Xun as magistrate some three years after winning his presented scholar degree in 1625. The account he kept of his time as magistrate there, published the year he was promoted to the post of censor in 1632, consists of summaries of 304 cases that he adjudicated or observed while in Xun. The title of his book, *Xunci* (*Xun judgments*), employs an archaic character (not the same *xun* as the name of the county) meaning to level a field after ploughing. To reflect Zhang's abiding concern to use the law to achieve fairness among litigants, I translate the title literally as *Level-Field Judgments*.[56] The roughness of the engraved woodblocks and the extensive use of nonstandard characters suggest that *Level-Field Judgments* was published locally in Xun county, languishing there as a little-known masterpiece of Ming legal history. The relative insignificance of the place suits our purpose, however, for it means that Zhang's cases reflect the conflicts, disappointments, and prices that colored the lives of ordinary people at the bottom of the Ming social world: potters, boatmen, gamblers, swindlers, and conmen out in the back of beyond. This was where many people of the Ming lived, far from the urban centers yielding most of the descriptive material we have regarding prices.

The recurring references to prices in Zhang's case summaries—the dated cases fall between 1627 and 1631—is hardly surprising. Few would put themselves in front of a magistrate were money not involved. When prices are mentioned, they are denominated in both currencies, generally in small amounts. The smallest sum Zhang reports is a theft of 200 coppers. Despite the judicial fact that theft was punishable by decapitation, Zhang regarded 200 coppers as so paltry that he let the thief off with a beating.[57] Most of the debts he settled were on the order of 1,200 to 1,500 coppers. The outlier is a case of an inveterate gambler who lost 2,500 coppers in one evening and turned out not to have a single coin in his pocket. This situation suggests that among the poor of Xun county, 2,500 coppers was a hefty sum, well beyond what an ordinary person could be expected to raise in ready money.[58]

Most who found themselves before the county magistrate dealt in sums and prices greater than these.[59] Some of these sums were for lost money, such as the traveler who lost 3,700 coppers that he was carrying to cover his expenses while on the road. Some were unpaid loans: the largest debt that I have noticed being a long-overdue debt of 7 taels. Other sums were for property. For example, Zhang heard several cases involving mules, one in which a man paid a down payment of 1,000 coppers on a mule priced at 5,000 coppers but failed to make the rest of the payment, and another in which the asking price of 8 taels was pushed down to 3 when the case went to court. Disputed land sales were also occasions for mentioning prices. In one case, someone spent 6,000 coppers to buy 12 *mu* (one *mu* was roughly equivalent to a sixth of an acre, or roughly 2½ tennis courts). In another, Zhang set the standard local price much lower, at 2½ taels per *mu*.

Zhang's casebook also reveals the prices paid for people. Concubines ranged in price from a meager 3 taels to a more generous 14,000 coppers. (Maintaining a mistress cost 600 coppers a month.) Brides were more expensive. The lowest price for a wife I have found in Zhang's casebook is 5,000 coppers, roughly 7 taels. In a complicated case, a man fleeing a famine in a neighboring county agreed to buy a young woman from her father for 14,000 coppers, the equivalent of about 20 taels. Not having the cash on hand, the man sold 12 *mu* of land (roughly two acres) for 6,000 coppers, paid that to the woman's father, then left the county without paying the remainder, which is why the sale came to the magistrate's attention.[60] Although Zhang Kentang provides no evidence, it bears noting that just as a bride could be bought, so could a groom. This was not a customary practice, nor a highly regarded one, as the term for the young man so purchased was *zhuixu*, literally, "useless son-in-law," for it was not the bride who was paying but her father. This arrangement was a way for a man to find someone to marry his daughter and then move into his family, a pattern known as uxorilocal (rather than patrilocal) marriage. The buyer thereby secured a male to labor for him, produce grandchildren, and perform other services. Legal historian Niida Noboru reprints a *zhuixu* contract from Huizhou dated 1593 in which the duties of the live-in son-in-law are specified: raising his children in his father-in-law's family, planting the family's fields, caring for his father-in-law in his old age, and never uttering a word of complaint. His price, which the contract politely phrases as "a gift of silver," was 15 taels.[61]

Zhang's highest prices are not for mules, land, or brides, but for attending funerals. A mourner was expected to present a cash gift upon arriving at a funeral. Zhang sometimes took advantage of this practice to arrange compensation for injured parties, assigning a gift in a specific amount to be paid by someone he regarded as prejudicially implicated in the suffering or death of the deceased. In one case, Zhang ordered a man who had wronged the deceased to give a gift of 3 taels at his funeral, a penalty significantly above what one was expected to give at a funeral. But he could and did impose even higher funeral gifts. In one case, the informal fine was 24 taels of silver; in another, 20,000 coppers, its rough equivalent.[62] These were substantial sums and may have represented something like an upper limit on the amount of money a moderately prosperous household in Xun county could be expected to raise on short notice.

Perhaps a better indicator of how much financial burden an ordinary person could carry are the unpaid debts Zhang had to sort out. In one case, a boatman owed a grain dealer 45 taels as a result of losing his wheat shipment when his barge tipped in a storm, but try as he did to pay down the debt, many years later he still owed the grain merchant over 27 taels. In another, a peddler left unpaid at the time of his death four IOUs for amounts from 1¼ taels to 8¾ taels, amounting to a total debt of 19½ taels. If we accept Las Cortes's estimate that a person's total assets might range between 20 and 30 taels, then these two cases confirm this range as the upper limit of financial capacity that ordinary people might reach, though with difficulty.

The Cost of Living in the Wanli Price Regime

From the fragments that writers such as Zhang Kentang, Adriano de las Cortes, and others have left us, it is just possible to reconstruct an approximation of the cost of living in the Wanli era. Following the practice that historians of Europe developed by estimating a "basket of goods" that a household had to purchase in order to survive and reproduce itself, I will now put together prices I have found to determine what a household had to spend to "guarantee a certain level of utility or welfare," in the language of price historians.[63]

In his study of the industrial revolution in Europe, Robert Allen has calculated the costs of living per adult male in early modern England in two baskets, a subsistence basket (what it cost a family to survive at a basic level) and a respectable basket (what it cost to maintain a "respectable" level of welfare).[64] The notion of subsistence varies from culture to culture, and certainly Las Cortes was amazed from his European point of view at how meager a subsistence basket in Guangdong could be—"a bit of bad rice and a few leaves of mustard greens," in his melancholy description—but that will not distract us here. Either one ate enough, wore enough, and had sufficient shelter to survive, or one did not. Developing a definition of what constitutes a "respectable" level of consumption is as dependent on cultural signals as on price signals, though here we are assisted by Wanli-era writers who, feeling somewhat embattled by the economic and cultural changes they were experiencing, took an interest in what was needed to maintain respectability.

Let us begin by establishing a basic food requirement. Allen began from the assumption that the average European adult male at the time was 165 centimeters tall and weighed 54 kilograms, and that such a person required a minimum of 44 grams of protein per day to survive, and something closer to 70 grams a day to perform heavy labor. To meet that standard, Allen set a minimum food intake of 1,900 calories a day, then created a subsistence basket for the average European adult male by combining oats (155 kg per year), beans and peas (20 kg), meat (5 kg), and butter or oil (3 kg), which adds up to an intake of 1,938 calories. To this food basket he added soap, cloth, candles, lamp oil, and fuel. This exercise does not easily transfer to Ming China, as we lack the data that Allen is able to extract for the English case. Studies of nineteenth-century adult males suggest that Chinese at that time were about 2 centimeters shorter than Allen's estimate for English males.[65] This difference allows us to postulate slightly lower caloric and protein requirements, though by how much is beyond my competence to determine.

To construct a basket of food for a person of the Ming, we can begin from a different point, and that is the handful of records by managerial landlords of the costs of supporting agricultural laborers. According to one such Wanli-era landlord from Huzhou prefecture on the Yangzi

delta, directly northwest of Chen Qide's Tongxiang county, a field laborer had to be paid an annual cash wage of 2.2 taels and a food allotment of 2.9 taels.[66] Assuming that he also provided accommodation at
no extra charge, the total of 5.1 taels may be taken to approximate a
subsistence basket for food and other basic costs. Another landlord
surnamed Shen from the same prefecture gives higher numbers. He
reckons that a full-time laborer cost him 13 taels a year: a wage of 5½ taels,
a grain allowance of 5½ taels (to buy 5½ hectoliters of rice), 1⅕ taels
for firewood and liquor, 1 tael for other foods, and ⅕ tael for tools.[67]
These prices come from late in the Chongzhen era, when grain prices
were double what they were in the Wanli era. Halve his grain price and
leave out the tools, and the laborer's food costs come to 5.15 taels. A
generation later, Zhang Lixiang, a native of Chen Qide's county of
Tongxiang who was responsible for preserving Shen's advice to farmers,
enlarged on what Shen recorded. Zhang kept the grain allowance but
itemized the other food costs: 73 catties of meat (at 2 cents a catty, 1.46
taels), 213 blocks of tofu (at a copper per block, roughly 30 cents), 273
ladles of liquor (valued at 0.82 taels), plus 2.6 taels for oil and firewood,
for a total of 5⅕ taels.[68] Add the discounted price of rice, and the total
cost of living comes in at 7 cents short of 8 taels.

The standard grain ration in the Ming for soldiers or famine victims
was one liter of grain a day, or 3⅔ hectoliters a year. This allowance
is considerably below Shen's estimate of 5½ hectoliters.[69] Perhaps this
volume of rice was necessary to attract workers in a competitive labor
market. Quite by chance, Las Cortes confirms Shen's per-person cost
of 5½ taels when he reports that the soldiers who held him and his fellow shipwreck victims captive were given 1½ cents per person per day
to feed them, which when annualized comes to 5½ taels.[70] In the case
of the agricultural laborer, what may account for the higher rate is that
he worked not just for his own grain allowance but for that for his
family. Since 5½ hectoliters is almost exactly one and a half times the
basic allowance of 3⅔ hectoliters, it would appear that he was being
given an additional half portion, whether for a wife or children, in
acknowledgment that he needed more than his own food minimum to
survive.

To enlarge the basket of goods to meet the needs of a wife and two children, I suggest doubling Zhang Lixiang's figures for grain and tofu for a single laborer, leaving the other amounts as they are, and adding 1 tael for eggs and vegetables. This formula produces a subsistence basket of food costing 12 taels. Las Cortes indirectly verifies this estimate when he reports that his hard-working tailor supported his family on 11 taels. To then estimate the cost of a basket of goods of a family above that of a poor laborer, we might increase what was spent on grain, oil, and firewood by 50 percent and double everything else to bring consumption up to a more comfortable level. If those assumptions hold, a respectable basket cost 18 taels.

But that is not the end of our calculations. To the cost of food we need to add other basic costs such as clothing and shelter. A simple suit of clothes in the fifteenth century cost 35 cents.[71] If the respectable family purchased one suit of clothing for each adult and child a year, and a child's suit cost half an adult's, this would add up to 1.4 taels per year, increasing the respectable basket to 19.4 taels. If we add half that amount to the subsistence basket, on the understanding that the poor wove and tailored most of their own clothes, that basket grows to 12.7 taels. To these budgets we need to add a few items of house and kitchenware. Relying on the prices that Hai Rui set for furnishing official residences in table 2.1 (in appendix C, "Tables for Reference"), I propose that we approximate what a family might spend in a year to maintain its household operations by taking a quarter of the value of the housewares and kitchenware. For the subsistence family, I would use Hai's prices for the lower rank (the right-hand column), which is 32 cents. For a respectable family, I would move to the middle column, which gives a figure of 57 cents. With this addition, our new totals are 13 taels for the subsistence family and 20 taels for the respectable family.

Finally, we need to add the cost of rent. A 1607 document from Nanjing records an official rental rate per "bay" (the distance between two beams, producing an area roughly 4.4 × 5.7 m) of 3.6 taels, noting however that this rate could be discounted by up to two-thirds for tenants unable to pay that much.[72] Using the discounted Nanjing rate of 1.2 taels per bay as a base, I suggest assigning the respectable family a three-bay house at 3.6 taels. Of course, rental rates could run much higher. An official notes in his autobiography that when he was a twenty-three-year-old preparing for the examinations in 1598, he was paying 10 taels a year for room and board.[73] But he was living well above the respectable level.

Summarizing this exercise, the annual cost of living of a respectable family in the Wanli era was just over 23 taels. For a family living close to subsistence, it was just over 14 taels. These estimates are loosely corroborated by an anecdote from coastal Fujian in 1595. On a visit to a shrine to the great Song Neo-Confucian philosopher Zhu Xi to conduct sacrifices in his honor, a prefect discovered that two of Zhu's descendants employed to care for the shrine were living in utter poverty. They supported themselves by growing lichees and vegetables on the shrine's property, which yielded an annual income of 10,000 coppers. At an exchange rate of 800 coppers per tael, the income measured in silver was 12½ taels. This was the amount on which they had to feed and clothe themselves (they were not charged rent). Desiring to improve their situation, the prefect organized a small fund-raising project to buy land that would yield a rent sufficient to support the shrine and its caretakers. The land produced an income of 30 hectoliters of rice, of which 5 hectoliters were to be set aside to cover repairs to the building and the other 25 given as income to the caretakers.[74] At a price of half a tael per hectoliter, the grain the caretakers received was worth 12½ taels, doubling their income to 25 taels. This provision thus raised them from a subsistence household to a respectable household, roughly confirming the costs of living estimate we just made.

Income

Income in the Ming period is exceedingly difficult to estimate, for three fairly simple reasons. First, while many sold their labor to survive, the majority did not earn wages. They worked as agriculturalists, growing their own food and exchanging what they produced for what they needed within a barter economy. The economic activities of most households thus lay beyond the reach of cash. Second, given that family budgets are nonexistent, it is impossible to reconstruct what portion of household income was earned through wages when wages were earned.

Those who sold their labor did so mainly to supplement household income rather than to earn it in full. So while a miserable wage might not meet the costs of even the most meagre basket, it made complete sense in helping to sustain a household when it was pooled with the income of other members. Third, the documents touching on wages that have survived are challenging to interpret, as we are about to see. Recorded wage rates do not necessarily reflect what the employer paid or the employee received.[75] Food was understood as an important component of a wage, for example, yet rarely was that component recorded.

Despite these difficulties, a few observations can be made about laborers' wages. Recurring references in Ming sources agree that the standard minimum wage throughout the period was 3 silver cents a day, or 24–25 copper coins. An official writing to the emperor at the end of the fifteenth century notes that laborers hired for military infrastructure projects were paid 1.1 to 1.2 taels per month, which amounts to a rate of 3 cents a day. This was clearly a minimum, for he goes on to deplore this wage. "Even when we sell them grain at a discount, it is not enough to feed them," he writes. In consequence, many laborers, "exhausted by the debts they accumulate, flee and disappear."[76] Writing over a century later, Xu Guangqi reported in a memorial of 1619 to Emperor Wanli that "a poor day laborer in the capital can make 24 or 25 coppers a day, which is barely enough to keep a man alive" and utterly insufficient to cover the cost of winter clothing.[77] Three cents is the amount Shen Bang records his office paying porters at the imperial family shrine in Beijing and cleaners hired for the day when girls were selected for the imperial harem.[78] Records from the Jingdezhen porcelain industry also record daily wages in this range: 3 cents plus 5 coppers for a highly skilled potter, half a cent less for workers preparing the cobalt, in both cases supplemented by a grain allotment.[79] Three cents is also what an official in Nanjing in 1635 paid a beautiful young man to spend the day with him cross-dressed clandestinely as a woman.[80] Of course, wages could go higher. Four cents was the daily rate for silk weavers, raised to 6 if they were using their own loom.[81] Six cents was the top daily wage in Shen Bang's accounts, paid to drummers and carpenters—with the curious exception of stove lighters (stove lighting was considered an almost magical art), who received 10 cents. In the closing years of the dynasty, the standard daily wage had risen to 4 cents, according to the diary of Ming loyalist Qi Biaojia, who paid local militiamen this rate.[82]

Not everyone earned the 3-cent minimum. An inscription commemorating the restoration of a bridge in Fujian in 1466 gives a daily wage for the workers of 2 cents "plus seven-tenths of a liter of grain a day, as well as vegetables occasionally and meat every other day." Although recorded in silver, the inscription reveals that these wages "were all paid in copper cash."[83] A century and a half later, Adriano de las Cortes reports that the lowest class of laborer in south China earned 2½ cents per day, or at most, 20 coppers.[84] Postulating a work year of 340 days, a daily wage of 3 cents / 24 coppers provided an annual income of 10.2 taels. Double that rate for a weaver with his own loom, and the annual salary is 20.4 taels.

It turns out that the most abundant set of annual wages is to be found in local gazetteers. They are there because local administrations were required to post their employees' wages, less because they were payments actually made than because they were derivatives of the imposition of corvée labor early in the Ming, and therefore fiscal data. These "fiscal wages" were devised and calculated in the later decades of the fifteenth century in the context of commuting corvée obligations into what was termed "labor service," which imposed on the taxpayer a payment in silver in lieu of personally performing the labor.[85] Since local offices then used this silver to hire substitutes, it is not unreasonable to assume that the commutations were calculated in relation to the actual cost of labor.[86] If fiscal wages have some referential value, it is because just as there were fair prices, so too there were fair wages.[87]

A survey of fiscal wage records in twenty-six randomly selected county gazetteers published between the years 1540 and 1630 reveals that half these wages were 5 taels or less, and three-quarters were 12 taels or less. Annual wages for the least skilled, from night criers to porters, ranged from 3⅔ to 5⅔ taels, not a wage on which one could survive and therefore better understood as a part-time wage.[88] Annual wages for those of moderate or more sought-after skills, such as scribes or mounted messengers, ranged from 8 to 11 taels.[89] Once we move up into the wages for men working in more responsible positions, such as a militia captain or a postal station master, the range is between 14 taels and 22 taels.[90] These wages coincide fairly closely with evidence that Minister of War Liang Tingdong included in a report on illegal foreign trade to Emperor Chongzhen in 1630, that Fujianese who took jobs as sailors on ships heading out into the South China Sea could expect to earn between 20 and 30 taels a year, a range regarded as a good wage.[91]

If we recall the earlier estimates of the cost of living (just over 14 taels for a family living close to subsistence, and just over 23 taels for a respectable family) and compare these with these wage data (a poor wage between 5 and 12 taels, a respectable wage between 14 and 22 taels), it becomes plausible to propose that wages were sufficient to meet the costs of living of Ming households.

A simple way to peer into wages further up the social scale is to turn to another official source, which is the pay scale for ranked officials working in the bureaucracy (see table 2.5 in this chapter). Officials were ranked into nine levels, each rank further subdivided into higher and lower, and were paid according to the resulting eighteen levels. In the Wanli era, the lower two-thirds of this scale ranged from 19.52 taels for a prison warden (rank 9b) to 66.916 taels for the second in command of a provincial commission (3b). Above that level, wages leapt to 88.84 taels for a vice minister (rank 3a), 152.176 for a censor in chief (rank 2a), and soared an impressive 265.511 taels for a grand secretary (rank 1a).[92] Although these are handsome figures, historian Ray Huang has noted they were "clearly inadequate in terms of the standard of living in the late Ming," which is why so many turned to gifts and bribes.[93]

Prices among the Wealthy

Costs of living depended on how one's social position determined how one had to live as well as what one could afford.[94] A rich man's bed could be a hundred times more expensive than a poor man's.[95] And whereas a poor man in Beijing could get by on 170 coppers a month (about 2½ silver cents), a rich man might have to spend 25 silver taels a month for room and board, again a ratio of roughly a hundred to one.[96] As one

TABLE 2.5. Salaries of Ranked Officials in Silver Taels, 1567

Rank		Representative Post at That Rank		Salary
1a	正一品	grand secretary	大學士	265.51
1b	從一品	minister	尚書	183.84
2a	正二品	censor-in-chief	御史	152.18
2b	從二品	provincial administration commissioner	布政使	120.51
3a	正三品	vice-minister	士郎	88.84
3b	從三品	administration vice-commissioner	參政	66.916
4a	正四品	prefect	知府	62.044
4b	從四品	National Academy chancellor	太學祭酒	54.736
5a	正五品	Hanlin academician	翰林學士	49.846
5b	從五品	subprefectural magistrate	知州	37.684
6a	正六品	bureau secretary	主事	35.46
6b	從六品	provincial registrar	經歷	29.084
7a	正七品	county magistrate	知縣	27.49
7b	從七品	supervising secretary	給事中	25.896
8a	正八品	county vice-magistrate	縣丞	24.302
8b	從八品	National Academy instructor	助教	22.708
9a	正九品	county assistant magistrate	主簿	21.114
9b	從九品	prison warden	司御	19.52

Source: Da Ming huidian, 39.1b-7b.

minister complained to Emperor Wanli, officials in Beijing easily spent as much as 4 or 5 taels a month, though the target of his complaint was not high rents but extravagant lifestyles.[97] These examples are simply intended to show that the rich lived in one price register, the poor in another, and the gulf between them grew wider as the commercial economy bit more deeply into the fabric of social life. China's commodity economy did not originate in the Ming, but if this period stands out from earlier eras, it is in the sheer number of people who were obliged to rely on commercial relations to survive, creating a situation in which prices served vividly to mark the gap between rich and poor. The Ming was thus a period for which there is abundant evidence of both great prosperity and great distress, and prices enable us to tell that story.

Thus far this discussion has focused on assessing what it cost ordinary people to survive. Let us now consider briefly what the wealthy spent, for this too can help throw light on the matrix of prices within which the people of the Ming lived. Our guide to the wealthy will once again be the attentive Jesuit observer Adriano de las Cortes. Although he lived among the poor, Las Cortes was curious about the officials with whom he came into contact and the world of wealth from which they came. As he had no personal entry into the lives of such people, he was not in the position to break down their budgets as he was the budgets of the poor. Still, he was alert to whatever he could glean and made a few notes in this regard. He observes that landownership was critical for maintaining a wealthy household. The poor might own a tiny plot to grow food for their family, but the rich owned large tracts of agricultural land. This land they either managed themselves using hired labor or rented out to tenants. Owning land on a large scale also tended to entail the provision of animal protein, whether by raising fish in ponds or keeping flocks of ducks and geese. According to Las Cortes, wealthy households also owned four to six water buffalo, a large number of pigs, and a dozen hens.

Las Cortes must have visited someone in this economic class, for he knew that their homes were well furnished. He writes of being amazed by the extraordinary number of skillfully worked chairs and tables in a single room, "as many as 25, 26, and even 40," as he phrased it, which surprised him because it was far beyond what one would find in a European room. He was also impressed with the quality of their bedding, fabrics, and boots. What further caught his attention was that "the wealth of rich Chinese also includes several slaves, men and women, who come from their lands and who were sold to them as slaves for life by their indigent parents, even when the latter are not in extreme necessity." Las Cortes was surprised and appalled to learn that slaves were cheap, although he does note by way of mitigation that servants in bondage "can redeem themselves for the price for which they were bought, and that their masters treat them well, yet in all matters they serve them like veritable slaves."[98] He provides a few prices. Fifteen-year-olds are worth from 1.2 to 2 taels, in contrast to a pig, which at full weight is worth 4 to 5 taels. Curiously, this is the same contrast that Chen Qide deployed in his famine memoir, albeit under different conditions. Nonetheless, this may be counted as one effect of the prosperity of the Wanli era: that some people became owners and others became owned. As for what the wealthy paid for a bride, Las Cortes asserts that among "the wealthy and respectable Chinese, . . . the bride price will be at least 500 taels, even 1,000, though very rarely, and solely among those who are really powerful and of the highest rank."[99]

Given their costs of living, the wealthy conducted their lives mostly in silver while the poor conducted theirs in copper. Las Cortes comments that a wealthy household had "a little silver" on hand to make purchases. He does not say how much, but another source puts a number on it. When the house of the Jesuit missionaries in Nanjing was raided in 1617 in an attempt to close down the Christian mission in that city, the inventory of their possessions that was drawn up after the arrest reveals that there was 17.6 taels of silver on hand, which may represent what a reasonably wealthy household had at home to manage daily financial dealings.[100] As for the poor, their cash on hand might have been more like the coppers a carefree man took from his pocket whenever he came home and tossed into a jar by the door just to ensure that he had enough loose change on hand should he need to buy some liquor to entertain unexpected guests.[101] This man kept no silver at home, living as he did in a world in which silver was not needed for everyday purchases.

Las Cortes concludes his survey of wealthy Chinese with the remarkable declaration that they were not as wealthy as their counterparts in Italy. From his perspective, the wealth of what Chinese called "thousand-tael families" did not match the wealth of a rich European household. He allowed that some of these families might have assets that brought them up to two or three thousand taels, though he notes that "even if one were to offer a more generous valuation, it would be difficult, for most of the rich, to get them up to 3,000 taels." Without diminishing the value of his observation, it may well be that Las Cortes was too much the outsider to know how to interpret local signs of wealth. He does acknowledge that he was not describing the richest families of the Ming, only estimating "the wealth of people who count as rich in China but are not merchants." He sensed that those whom trade enriched were in an economic world apart from the landowning gentry, and to that world he had no access.

The lack of family accounts among the surviving documents of the Ming means that we have almost no way to enter the world of prices that the wealthiest inhabited. The one exception I have encountered is the prices for artworks. The prices of art and antiques soared during the Wanli era, so much so that some collectors kept notes, and many observers took notes, of the sometimes extraordinary prices that the wealthiest paid for works of art and other luxury goods. I will close this survey of the Wanli price regime by making a brief foray into the art market.

Prices in the Luxury Economy

Li Rihua was at home in the city of Jiaxing, just twenty-five kilometers from Chen Qide's county of Tongxiang, on 11 August 1612 when an antiques dealer named Sun from further up the Yangzi delta moored his barge along the riverbank below his home and sent a servant up to let Li know he had arrived. Another dealer happened to be visiting Li at the time, and the two of them sauntered down to the barge to see what dealer Sun had on offer. Delighted that they deigned to come onto his barge, Sun brought out item after item in the hope of making a sale. He started by showing them twenty-four decorative bronzes allegedly from the Imperial Household, including two rectangular incense burners by a famous craftsman and a footed tripod cast during the Xuande era (1426–35), renowned for the quality of imperial manufactures. Sun had one other Xuande-era bronze in his collection, a small incense burner, to which he could attach no provenance. He also showed them an ancient cup fashioned from rhinoceros horn, a porcelain oil lamp made at one of the imperially commissioned kilns during the Chenghua era (1465–87), and two fine folding chairs, their backs inlaid with expensive Dali stone, which he claimed had come from a venerable Buddhist monastery.[102]

Dealer Sun's prize piece was a porcelain cup in the shape of a magnolia blossom, which he claimed was produced at an imperial kiln of the Xuande era. That provenance was worth its weight in silver, and Li did not doubt it. The cup, he writes, "was wonderfully archaic and unforced, the inside lustrous and white, the outside glazed in a pale purple with an interlocking flower design underneath." He was charmed until Sun told him the price: 40 taels of silver. Li was appalled. "And so it is," he sighed, "that we have now arrived at the point when the product of a tile kiln is costlier than gold or jade!" Li's reference to tile kilns, a debased site of industrial production, was meant not to disparage the cup but to accentuate how the order of value had so changed that demand could push the price of a piece of manufacture above its natural position in a fair price regime. By losing its appropriate value, the cup had gone from being a delightful object to a wealth fetish—which is why it is of interest to us, even if it wasn't to Li.

The luxury economy was not a peculiarity of the Ming era. What sets the Ming apart from earlier eras is the breadth of participation in that economy. One impact of that breadth was that by the time Li Rihua was recording his adventures in the luxury trade, demand was driving prices to extraordinary levels. Not only were wealthy buyers competing for objects at the top of the market, but an even larger number of middling buyers clustered below them, bidding on what they could afford. To grow to this scale, the luxury market had to have been bolstered by the larger economy in which more people had greater liquid wealth than ever before and chose to invest more of that wealth in luxuries, whether to serve as tokens of status or as stores of wealth.[103]

One writer who tracked the escalation in luxury consumption in the Wanli era was Li Rihua's brother-in-law, Shen Defu. Shen includes a long treatment of the subject in *Wanli yehuo bian* (Private gleanings from the Wanli era), which he published in 1617 just as that era was about to come to an end. According to Shen, it was in the middle decades of the sixteenth century that rich people started to amass enough wealth to get above their stations, building gardens and assembling private opera troupes as though such projects were no longer privileges only the court could enjoy. More to the point, the newly wealthy spent that wealth "amassing collections of rare antiques and collectibles over several generations without tiring of searching everywhere for them." Shen names Li's mentor, Xiang Yuanbian, among the prominent collectors who "did not stint on paying huge prices to buy such things" and whose reputations spread throughout Jiangnan. As a result, according to another participant-observer, the masterpieces of the painting tradition "were going round in circles and being traded back and forth."[104]

The volatility of the art market became a point of particular anxiety for elite collectors such as Xiang Yuanbian, who were accustomed to buying and selling their works within a fairly closed circle. Works of art were now moving outside the circle of gentry connoisseurs who claimed to prize these artworks not as investments but as embodiments of the loftiest values of high culture, going instead to men of wealth who could outbid gentry buyers. As the prices soared, entry to this market was restricted to those who could afford to pay prices in the thousands of taels of silver, which was beyond the capacity of many if not most gentry collectors. If the gentry could not outbid these new buyers, they could at least disdain them as fools with only money and no taste. Xie Zhaozhe, another caustic observer of the Wanli-era cultural scene but from the vantage point of Fujian rather than the Yangzi delta, observed that nine out of every ten buyers "now go about in the fine silk trousers of the nouveaux riches. Their silver accumulates, and they can find no place to put it. When they hear that an important piece of calligraphy or painting is available, they buy whatever is sitting on a shelf or hanging on a wall."[105] Shen Defu worried further that the new high prices would lead to social calamity.[106] Li Rihua agreed with his brother-in-law. "Fundamentally, calligraphy and painting are occasions for savoring elegance," Li Rihua writes. "What do those who entangle themselves politically to the point of gambling with their lives, who curry favor to the extent of courting disaster, have to do with things that are pure and precious? Is it not because they are simply making use of them as capital to gain wealth and power?" Li's use of the term "capital," *zi*, snags the attention of the modern Chinese reader, albeit anachronistically. But when money needs to be converted into objects that can boost its value, then as now artworks are regarded as sound investments.

To characterize prices in the Wanli-era art market was even beyond those who participated in that market. As artist Tang Zhixie wrote late in the dynasty, the art market was peculiar in that prices did not converge or find a level. You might be able to come up with a fair estimate for the price of a Ming work by considering the fineness of the craftsmanship and the fame of the artist, he observed, but pre-Ming works were far too volatile to submit to rational pricing. After regaling his readers with a few stories of absurdly high prices, Tang concludes that "stories like this are unending, so the blind should stop asking about the price of paintings."[107] If Tang had no clear analysis of the art market, it was because prices in the Wanli era were so far out of line with art prices in earlier times. It was still possible in the mid-sixteenth century to acquire significant masterworks for not unreasonable prices, but once into the Wanli era, prized artworks—of the Yuan period, for example—were said to have increased tenfold.[108]

To give some shape to the Wanli-era art market, I have grouped the 112 art prices that I have been able to discover into thirds.[109] The median price in the bottom third is only 2½ taels, indicating that the art world allowed entry to those with comparatively little to spend. The median price of the middle third rises to 30 taels, which sits cleanly above the annual cost of living of a respectable family. The median for the top third is 300 taels, quite out of reach of most people.

This analysis looks slightly different when focusing on a single collection rather than purchases all across the market, in this case the collection of Li Rihua's mentor Xiang Yuanbian, one of the leading collectors of the early Wanli era. Xiang kept close track of what he spent, often writing the price he paid on the label attached to the outside of a scroll. Sixty-nine such prices survive.[110] Dividing Xiang's collection into thirds reveals a pattern that differs significantly from the art market as a whole. The biggest difference is at the bottom end of Xiang's collection. The median for his lower third is 20 taels, eight times higher than the median of 2½ taels for the art market as a whole. The difference signals that Xiang did not bother buying works at the bottom end of the market. In fact, he records no price under 3 taels. The median price of Xiang's middle tier is 50 taels, notably higher than the 30 taels that marked the middle third of the art market. It seems that Xiang preferred to buy this higher level, which allowed him to acquire works that were not the conspicuous masterpieces that the nouveaux riches were after and that stayed within a range from 30 to 100 taels. When we get to Xiang's top third, which ranged from 100 taels to 1,000, the median price is 300 taels—exactly the same as the median price for the top third of all works of art. Only a few people collected at this level, and the prices paid there were the same for everyone whether you were Xiang or weren't.

Xiang Yuanbian was far from being the wealthiest man in the Wanli era. He was a member of the gentry, earnest in supporting its cultural ideals and so careful of his finances that his elder brother recollected that he could go into an agony of self-recrimination when he decided that he had paid too much for a piece of art. If Xiang assembled a significant collection of art that would shape elite Chinese taste for centuries into the future, it was because he was from a family that had the means to acquire the classics that portrayed that taste as well as the cultural capital to communicate that definition, and did so just before "unlicensed" buyers entered the market. When Xiang died in 1590, two-fifths of the way through the Wanli era, the competition for masterpieces was rising rapidly, forcing prices above even the highest he paid.

Art prices are not sufficient to describe the costs of living of the richest families of the Ming, but they do give a sense of the very different price regime that the wealthy inhabited. Adriano de las Cortes had no point of contact with this rarefied social circle and knew nothing about the prices at which they circulated their treasures. Nor did Chen Qide. Chen was aware that a local luxury market existed, for he notes that during the first wave of famine in 1641 "no one even stopped to ask" dealers about the prices of works of art and fine curios. Perhaps he bought a few items at the bottom end of the art market in the range of 2½ taels, but nothing grander. To the extent that the market for fine art and expensive antiques existed in his own time, it operated well beyond his financial capacity and possibly even his interest. He could dismiss art collecting as an indulgence of the morally lax rather than an instance of the culturally profound. In any case, in 1641, the Chen family was not buying art. If they had any silver on hand, they were spending it on food, not luxuries.

3

Silver, Prices, and Maritime Trade

MANY PEOPLE EXPERIENCED the Wanli era as a time of prosperity, innovation, and new ideas. Some historians have attributed this wave of prosperity and intellectual renaissance to the new networks of exchange stretching from the South China Sea westward to the Indian Ocean and Europe and eastward to the Americas. Foreigners were approaching China in ever greater numbers to purchase products—silks, porcelains, furniture—that were superior in workmanship and lower in price than similar manufactures made anywhere else, paying for their purchases with silver that was being mined in Japan, Mexico, and Peru. As the silver flowed into China, it has been argued, the economy grew, prices rose, society shifted, and new philosophies gained purchase. Buried inside this hypothesis is another, which is that when the flow of silver dwindled in the 1640s as production declined in the Americas and Japan sealed its borders, that constriction strangled the commercial edifice that the growth in money supply had induced, tipping China into economic crisis.[1]

The new presence of European traders in the Indian Ocean and the South China Sea drew Chinese into networks of trade that were broader, deeper, and more sustained than the trade links with Southeast Asia and the Indian Ocean before the sixteenth century. This change is unquestionable. But what place does that engagement have in the story of rising prices and economic crisis that Chen Qide relates? In his writings, the schoolmaster has not a word to say about the world beyond China, other than to count being born inside China as one of the ten fortunate things of his life.² Chen Qide was one of the many people of the Ming who were not conscious of being affected by the foreign trade that was enmeshing some Chinese into trade networks stretching beyond the shores of the Ming. People of this sort were embedded mainly in the agricultural economy and viewed foreign trade as a commerce in unobtainable luxuries that touched few and was alien to most. Within his world, nothing consumed as a daily necessity, even as an ordinary luxury, arrived as an import. The trade that carried Chinese goods out into the world was beyond the awareness of most people other than those who lived along the coast and engaged, usually surreptitiously, in maritime trade.³

Insulating the people from foreign trade was politically intended, for the Ming state monopolized the right to license foreign contact and foreign trade. It did so through what is known as the tribute system, a set of protocols by which foreign rulers presented gifts to the emperor in order to enjoy his favor and obtain for their envoys a limited right to trade while in the country. This state monopoly on foreign trade prevailed through the first half of the dynasty. The situation shifted in the middle of the Ming period as the maritime pathways into the world, disrupted after the fall of the Yuan Great State of the Mongols, resumed commercial traffic. In the second half of the dynasty, private merchant ships increasingly came and went from the southeast coast. The profits were large, as was the cost of entry. Building an oceangoing ship in Fujian around 1600 cost well over a thousand taels. And that was only the initial costs, for after it sailed, a ship had to be serviced annually and refitted for its next voyage to the tune of some 500 to 600 taels.⁴

Seventeenth-century essayist Tan Qian gives us a taste of mid-Ming maritime trading by recounting an imperial order to the Ministry of Revenue in 1555 for 100 catties of ambergris for the palace.⁵ A waxy substance produced in the digestive tract of sperm whales, ambergris was used as a perfume fixative. Known in Chinese as "Longyan incense" because it came from the Sumatran island of Longyan, it fetched an enormous price. Tan writes that a catty of ambergris was priced at 192 coins "of the gold money of that country," which he converts to 9,000 Chinese coppers, or about 12 silver taels. The Ministry of Revenue reported to the emperor that there was no ambergris to be had in Beijing, so the court ordered officials in coastal areas to buy what they could find. Officials in Guangdong were best placed to respond. At first they could only secure a small amount at the shocking price of 200 taels per catty, and even then it turned out not to be the real thing. Eventually they tracked down true ambergris in the possession of a foreigner named (in Chinese) Manabiede. He was in prison at the time, presumably for smuggling Southeast Asian products into Guangdong. He had less than a tenth of a catty of ambergris, but that was better than nothing, and it was sent to Beijing. Then another Southeast Asian merchant from an island transcribed in Chinese as Midishan came forward with 6 catties. The Midishan merchant's ambergris was sent to the palace the following year. Once word reached Sumatra that the Ming court wanted ambergris, supply was assured, though it had to pass through a certain legal ambiguity to reach its buyers. In the hope of not paying over the odds, the court capped the price of ambergris at 100 taels per catty.

The date Tan assigns to the ambergris story, 1555, was right in the midst of a period when the Ming state was enforcing the complete ban on maritime trade that Emperor Jiajing had imposed thirty years earlier. The only foreign commodities allowed to enter the country were goods brought by tribute missions. Tan Qian's story is evidence that the ban was both enforced (Manabiede was in prison) and not enforced (the merchant from Midishan was not in custody, even though he was not an official tribute envoy). In addition, the story reveals that local officials in Guangdong who wanted a Southeast Asian commodity knew how to find it. Even more curious is the reference to 192 coins of local currency. Tan Qian provides the conversion rate of nine strings of Chinese coppers. Chinese coins were in demand in Southeast Asia and would have exchanged for silver at a strong demand rate, which could have run as high as 600 coppers per silver tael, such that the 9,000 coppers Tan cites could have been worth as much as 15 taels. Divide the silver weight of 15 taels (560 g) into 192 coins, and you have a coin with a silver weight of 2.9 grams. As it happens, this is just under the nominal value of a real, a Spanish coin minted in Peru that came into wide circulation throughout Southeast Asia after Spain established its colony in

Manila in 1571, just a year before the child emperor Wanli was put on the throne.[6] The year 1555 is too early for a reference to Spanish coins in Sumatra, though as the Portuguese had already established trading posts in Macau and Malacca, this detail may reflect the early circulation in this region of European coins.

Dealing in the market for ambergris required more silver than most people could imagine, let alone raise. By the Wanli era, the flow of silver was reversing. Rather than leaving China to purchase foreign commodities, the precious metal began coming into China from the mines of Japan as well as the New World to pay for Chinese products in a flow that grew through the Wanli to Chongzhen reigns as foreign traders competed to move Chinese merchandise abroad. The flow is undocumented from the Chinese side, since the state did not take a role in the trade and the merchants who did held their account books in absolute secrecy. This lack of Chinese records has obliged historians to extrapolate from foreign sources. Richard von Glahn has assessed these extrapolations and estimated an average rate of import of 46,600 kilograms of silver a year through the last three decades of the sixteenth century, rising to an average of 116,000 kilograms a year through the next four decades. During the Wanli era, he estimates that 60 percent of that silver arrived from Japan to buy Chinese goods that Japanese merchants then reexported around the South China Sea, and this despite the Ming court's particular ban on trade with Japanese. The rest came mostly from the Spanish mines in South America, brought mainly by Spanish merchants to sustain their vast wholesale trade through Manila back to the Americas. The roughly fifty metric tons of silver that crossed the Pacific annually is certainly a lot. When set in the global context of the period, however, that weight of silver represented only 7½ percent of the total output of the Peruvian mines during the Wanli era.[7] The lion's share was being carried to Europe and elsewhere, most of it to remain in Europe but some to reach China later through other European carriers.

Did the silver flowing into the Ming so enlarge the supply of money in an economy of roughly a hundred million people that it forced prices to rise? Many historians have concluded on the basis of fairly spotty and unreliable price data that it did, though most economic historians have now backed away from that hypothesis. This chapter will side with the latter group to cast doubt on this assertion and suggest a different interpretation. We will start by reversing our gaze and approaching the subject of silver in international trade from the perspective of the English who in the late Ming were striving to ride the wave of silver to get themselves into the China trade. Then we will turn to China to examine the role of prices, first in the tribute system during the first half of the dynasty, then in the last half century of the Ming when Spanish, English, and Dutch merchants not only traded in commodities and silver but kept fairly reliable records of what it all cost. The chapter will end with a brief consideration of some contemporary Ming voices on the beneficial effects of the silver trade.

Overseas Trade

From the sixteenth century, silver was the principal medium of exchange, not just in China but around the world. Prices were set in silver, payments were made in silver, whether coined (as in Europe and the Americas) or unminted (as in Asia), and accounts were kept in its units. Silver's role as the global currency of this age is well recognized. Rather than repeat what is known, I shall approach the use of silver in the global trading regime of the seventeenth century by starting from an eccentric position at the opposite end of Eurasia and asking how English traders used silver to conduct trade with China. The eccentricity is not simply that England is not China. It is that England, as much of Europe, was a trading nation that used silver but did not have any significant domestic source of the metal. England's access to silver depended entirely on its participation in global trade networks and its capacity to move goods among markets. Our guiding assumption as we proceed is that we should not naturalize the use of silver as the global medium of exchange. Silver did not "flow" of its accord, as though it were water simply flowing downhill. It moved in the ways it did, not just to China but throughout the system, because those who controlled its supply used it strategically to optimize the benefits that they could derive from that use.[8]

On the last day of 1600, Elizabeth I authorized the formation of a new corporation known as the East India Company (EIC), granting its cofounders a monopoly on trade with Asia. The unstated impetus for this initiative was the need to compete with Spain and Portugal for access to Asian markets. The company grew slowly until 1620, when the English economy went into recession. The popular perception at the time was that the recession was due to a shortage of coins. Without coins to use, the argument went, buyers could not buy, and so prices collapsed. Many blamed the shortage on the East India Company for its policy of shipping English coins overseas to purchase commodities abroad. The EIC was indeed exporting bullion. Company statistics that James Mill accessed to write his *History of British India*, published in 1817, reveal that bullion exports did in fact grow through the mid-1610s, from £13,942 per annum in 1614 to £52,087 in 1616.[9] Vulnerable to both popular and royal opinion, the East India Company had to react. The spokesperson who stepped forward to defend the company on this issue was Thomas Mun. By the time he was appointed a director of the EIC in 1615, Mun had had considerable experience of foreign trade in the open port of Livorno, Italy, and was as confident a mercantilist (advocating that national wealth was best built through advantageous foreign trade) as you could find in the 1620s. His written response to the charge that the EIC was draining England of its silver appeared as *A Discourse of Trade, from England unto the East-Indies*. This pamphlet garnered much attention, going through two editions in its first year of publication (1621) and then finding a wider readership when Samuel Purchas included it four years later in his omnibus collection of texts on England's foreign relations and trade, *Purchas His Pilgrimes*.[10]

"The trade of merchandise," Mun declares in his opening sentence, is "the very touchstone of a kingdom's prosperity." This declaration rests on the argument that wealth is to be counted not by the volume of specie a country has on hand, but by the volume of goods and specie it has in circulation. So long as imports do not exceed exports, money that countries spend abroad "must return to them in treasure," and at a greater rate than the simple difference between the value of exports and the value of imports. To rebut the popular claim that "the gold, silver, and coin of Christendom, and particularly of this kingdom, is exhausted to buy unnecessary wares," Mun presents his readers with price data for six commodities that the EIC was buying in Asia: pepper, cloves, mace, nutmeg, indigo, and silk (see table 3.1 in appendix C, "Tables for Reference").

Mun makes his case for the benefit of trading in pepper, England's largest-volume import at the time, this way: Europe, he writes, buys 6 million pounds of pepper per annum. The cost of buying that amount of pepper wholesale in Aleppo, at a rate of 2 shillings per pound inclusive of all charges, is £600,000. To make the same purchase in the "East Indies"—Mun's indeterminate designation for the zone from India's Malabar Coast to Java—at only 2½ pence per pound amounts to £62,500, just over a tenth the Aleppo price, thereby reducing the retail price in London. When pepper came from "Turkey," meaning the Ottoman empire, the price in London at the best of times was 3 shillings 6 pence, whereas when it came from the "East Indies," the price fell to 1 shilling 8 pence, less than half the Turkish rate. Sometimes it even went as low as 1 shilling 4 pence. Instead of making consumers pay £70,000 for the 400,000 pounds of pepper that the EIC imported into London, the company was charging them only £33,333. Mun repeats this calculation for the five other imports, giving readers what the prices in London were "formerly" as compared to "latterly."[11] One of these imports is silk, though regrettably for our purposes, Persian rather than Chinese. It would be at least a century before China would replace Persia as the main source for Asian silk coming into Europe.

Mun's point is simple. By shifting its purchases to Asia, the company was able to lower prices on imports, acquiring these commodities for a fraction of what it would have had to pay in the Levant. The benefit to English consumers was lower prices, though Mun liked to point out that the trade furnished other economic benefits: paying income to EIC employees, providing support to their families when they were overseas (prorated according to the wage earner's length of service), distributing relief to widows and children in the event of their deaths, and donating funds to such philanthropic projects as the "repairing of churches, maintenance of some young scholars, relieving of many poor preachers of the Gospel yearly with good sums of money; and divers other acts of charity." In all these ways, Mun was confident, company trade benefited England and its people. "All the money which is sent forth in our ships doth procure an overplus of the said wares to the furtherance of trade from India hither, and after from hence to foreign parts again, to the great employment of the subjects and enriching of this realm, both in stock and treasure," meaning goods and cash.[12] Rather than draining England, Mun concludes, exporting bullion meant buying foreign commodities closer to their point of origin, thereby reducing the price for English consumers and keeping the profits in England.

It is striking that Mun chose to demonstrate his argument by using price data, and even more so that he focuses so single-mindedly on "treasure." In the language of the time, the word meant precious metals, mainly gold, silver, and coin. (English coins, pence as well as shillings, were mostly minted in silver; Mun's figures imply that only 2 percent of English coinage was cast in gold.) When Mun says "treasure," therefore, he is referring specifically to silver, which was the international medium of exchange in Stuart England just as it was in Ming China. What Mun's calculations demonstrate is that silver was as central to the concerns of English traders as it was of Chinese traders. Like China, England was not a major silver producer; unlike China, it did not produce commodities in as high demand as those the English bought in Asia. In neither case did the nonproduction of silver matter, however. What mattered was the capacity to use silver to gain entry into global trade networks. Whether you were an East India Company director sending silver and receiving imports, or a Ming merchant sending exports and receiving silver, you made sure that you positioned yourself to profit from the trade that silver lubricated. Without the extensive mining and refining of silver at multiple sites, including the Americas, Ming merchants would not have been able to sell into foreign markets as profitably as they did—just as, albeit from the opposite direction, without silver East India Company merchants could not have moved the commodities they traded as profitably as they did.

The price of silver when measured against gold was not constant throughout the system. In Ming China, it hovered around 1 tael of gold exchanging for 5 taels of silver.[13] Other economies priced gold more highly than China did. Europe exchanged gold for silver in the range of 1:12. Japan followed suit in the late years of the Ming, partly as a result of its expanded silver production, but partly also under the pressure of exchange rates elsewhere. As a result, as Richard von Glahn has explained, Ming China attracted silver and shed gold.[14] Chinese manufactures were desirable and reasonably priced, the arbitrage profits of dealing in silver increased the eagerness of foreign buyers to buy, and the exchange rate only heightened that eagerness.

We now shift to the view from China, starting in the phase when the Ming court controlled foreign trade through the tribute system, in effect monopolizing it for benefit of the Imperial Household—quite a different strategy from the one Elizabeth I endorsed when she gave the monopoly to rich London merchants.

Tribute and Trade

The institutional structure through which the Ming court managed its foreign relations before the emergence of the silver-based maritime trading networks has been called the tribute system in deference to its political functions. The system set the protocols for receiving envoys bringing tribute from foreign rulers to lay before the emperor, and for sending them home, duly honored and gifted. The system entailed considerable costs to the envoys, who were expected to bring gifts of suitable value to present to the emperor of China. It was also a burden on the Ming state, which undertook to cover all their expenses while in China and to provide return gifts of value equal to those presented to the emperor. It was an unequal ritual relationship that disadvantaged the tributary, but tributaries acceded to it on the expectation that the wealth their envoys carried out of China would exceed the value of the gifts they took to China.

The agency assigned to manage tribute relations at the levels of both policy and practice was the Ministry of Rites. However, because the management of gifts and imports generated revenue, the ministry regularly found itself in competition with the Imperial Household, which through the early decades of the dynasty took over the administration and collection of customs duties. The agency for managing custom duties, the Maritime Trade Supervisorate, was staffed by Imperial Household eunuchs. Envoys and traders had to go through the state monopoly this office represented and pay fees for the privilege of trading in China. The Ming started with one customs house at Taicang near the mouth of the Yangzi River. The unruliness of private traders in the Yangzi estuary persuaded Emperor Hongwu in 1374 to close Taicang and move customs collection to three customs houses down the coast. The Ningbo house was set up in Zhejiang province to receive Japanese envoys, the Quanzhou house in Fujian province to manage delegations from the Ryukyu Islands, and the Canton (Guangzhou) house in Guangdong to deal with ships from the "Western Ocean," as the maritime zone from the South China Sea westward to the Indian Ocean was called. The Ningbo house was shut down in consequence of a dispute between competing Japanese delegations in 1523, after which all trade with Japan was interdicted until the end of the dynasty. The Quanzhou house was later closed during the Jiajing ban on foreign trade, leaving only the Canton station to manage the traders and commodities coming from abroad.[15] When the tomb of eunuch Wei Juan, who headed the Canton Maritime Trade Office from 1476 to 1488, was excavated in 1964, archaeologists found three foreign silver coins, two from Bengal and one from Venice, which shows that, even in the fifteenth century, foreigners were bringing foreign cash to buy Chinese goods.[16]

Foreign envoys entered the country to fulfil diplomatic tasks, but they were also there for the economic tasks of buying and selling. Everyone understood the arrangement, even if more conservative officials would rather this were not the case. In general, diplomatic requirements prevailed over trade concerns. In 1447, for example, the court made the sale of blue-and-white porcelains to foreign envoys a capital crime for Chinese in order that the court keep for itself the privilege of distributing highly prized porcelains to foreigners as political gifts.[17] The tribute that Chinese law required embassies to bring into the country to present to the emperor were not so much free "gifts" as goods that the court required the embassy to present and for which it paid current prices.

The official record of foreign embassies, spottily preserved in the *Collected Statutes of the Ming Great State*, repeatedly states that envoys were "paid the prices" of the goods they brought into the country.[18]

Determining what prices both sides considered fair could pose a challenge, as the Ministry of Rites discovered in 1526 when an embassy from Lumi, a port-state in the Indian Ocean, showed up hoping to do business. Tribute regulations required Lumi to present the emperor with lions, rhinoceroses, and a bountiful collection of luxury products of the region, including jade and gems, and it did. The embassy also, though, brought a large number of iron pans, which were not on the list of the tribute items Lumi was required to present. These pans created a minor crisis for the Ministry of Rites, which did not know what price to put on these objects. Worse, new regulations brought in some three decades earlier to limit tribute gifts, and thereby control the scale of the bill the court had to pay, forbade nonregulation goods, such as iron pans, from being presented to the throne. The ministry advised Emperor Jiajing to take the indulgent view, to treat this as a case of the envoys' being unfamiliar with the new rules and accept the gift. When the head of the embassy then memorialized Jiajing to point out that the total value of the goods he had brought was worth over 2,300 taels and reminded him that the return trip took seven years (both were exaggerations designed to drive up the tab the court was supposed to cover), Jiajing asked the Ministry of Rites what he should do. In a joint response, a ministry secretary and a capital censor huffed that this delegation was clearly in China not to affirm Lumi's submission as a tributary but to pursue "the business of merchants setting profit margins." Having got that off their chests, they advised that it would be best for the emperor to accept the gifts rather than cause offense, but to make clear that goods in excess of what was required should not be presented in the future. Jiajing recompensed the envoys following the old protocol of matching the value of the tribute gifts (though no price for the pans was ever recorded), adding that they should follow the new protocol next time they came—which, as it turned out, was never.[19]

Prices thus mattered inside the tribute system just as they mattered outside the system as, increasingly through the fifteenth and sixteenth century, envoys offered goods for sale to Chinese merchants. This was an arrangement that the Ming state did not favor and only barely tolerated. When foreigners sold to Chinese, the policy, as the dynastic history puts it, was "to manage the foreigners' desire [to trade] while suppressing corrupt [Chinese] merchants."[20] One way to exert that management was to set prices. Whatever goods tribute envoys brought to sell, whether officially or on their own account, they should be "paid the price" at which they were valued on the open market. The goal was to avoid disputes and uphold China's moral authority by ensuring that "the exchange is fair to both parties."[21] To reduce ambiguity about fair pricing, Emperor Hongzhi at the end of the fifteenth century ordered the Imperial Household to draw up a list of "set prices based on realistic estimates." A prefatory note to this list, which has been preserved in the *Collected Statutes*, reminds the reader that previously foreign envoys were not permitted to engage in the private sale of goods, and that the commodities they might bring into China for private sale were subject to official confiscation if discovered. But blind eyes were turned and loopholes opened, and substantial volumes of foreign commodities flowed through the gaps.[22]

Despite headaches on both sides, the pressure to trade was unrelenting. The Ming produced goods that neighboring states throughout East and Southeast Asia wanted to buy, and Ming merchants were eager to sell them to these buyers. By the end of the fifteenth century the volume of illegal trade outweighed the volume of legal trade permitted under the rubric of tribute. Not all officials were content with this situation. Min Gui, supreme commander of Guangdong and Guangxi provinces, appealed to Emperor Hongzhi in 1493 to crack down on the huge number of foreign ships that were landing in China without reporting their arrival to officials and without any regard for the tribute schedule, which limited their arrival to every two or three years. The incoming traffic was beyond the capacities of Min's naval forces to monitor. And since customs revenues were partly streamed into military budgets, smuggling was at the expense of his budget. Hongzhi forwarded Min's request to the Ministry of Rites for an opinion. In his response, the minister was as unenthusiastic about the tribute system as he could be without going all the way to actually suggesting it be abandoned. A lax border policy in principle was not a good idea, he allowed, as it simply encouraged more foreign ships to come into Ming waters. On the other hand, an overly strict policy on foreign trade would strangle the flow, entailing severe economic loss for the region. The minister gently reminded the throne that "cherishing men from afar," the standard elliptical formula for tolerating foreigners while keeping them at arm's length, should go hand in hand with providing a sufficiency for the country. In other words, rather than curtail the trade, the emperor should let it continue. As duties were difficult to collect, the wisest course was to do nothing. Inaction on this score would allow the people of the coastal regions to gain some economic benefit from the trade without seriously compromising the state's budget or its security. The Hongzhi emperor was content with this response and made no changes.[23] Supreme Commander Min must have been furious.

This struggle between those who wanted to restrict trade in order to control it and those who wanted to allow trade in order to tax its benefits and spread the wealth intensified in the sixteenth century. Where the line of fissure lay had usually to do with perceptions of state security rather than any deep bias for or against trade. Increasing pressure coming from foreign ships arriving on the coast of Guangdong, among whom Portuguese began to be counted, provoked another round of noisy debate at court in 1514. When a provincial assistant administration commissioner in Guangdong launched the first salvo by seeking to impeach his immediate superior for his lenience, he invoked the very items that the Hongzhi emperor had okayed for informal sale: "The commodities circulating in south China, which come from foreigners from Malacca, Siam, and Java, are nothing but pepper, sappanwood, elephant tusks, tortoise shell, and such like," not "daily necessities such as cloth, silks, vegetables, and grain." This Confucian-tinged complaint about nonnecessities was a convenient way of disguising the real anxiety at work here: not the consumption habits of Chinese but the security threat posed by the prospect of "thousands of evil persons building huge ships, privately purchasing arms, sailing unhindered on the ocean, illicitly linking up with foreigners, and inflicting great harm on the region."[24] What regional officials regarded as Portuguese misconduct on the south coast compromised the policy debate that started in 1514. By the time the reigning emperor died in 1521, those at court opposed to allowing trade to expand had the upper hand. The Ningbo customs house was closed the following year, and three years later, in 1525, Emperor Jiajing banned any ship equipped with more than one mast from going to sea. The sea remained closed down to the end of his reign in 1567. Pressure from southern coastal Fujian, especially Quanzhou and Zhangzhou prefectures, to reopen the coast only grew. A sign that that pressure was having some effect was the elevation of the port of Moon Harbor (Yuegang) in Zhangzhou, which dominated late-Ming maritime trade into the South China Sea, to county status on 17 January 1567, inaugurating Haicheng county. When Emperor Jiajing died six days later, his ban on foreign trade could finally be reversed.

A generation later, as silver was moving in greater quantities through maritime trade networks, essayist Shen Defu could look back from the Wanli era and declare the ban on foreign trade to have been bad policy. It had denied the government valuable customs revenue, and it had deprived coastal residents of opportunities to enrich themselves, driving the wealthy families into cahoots with smugglers. Even so, the elite of Fujian province were split on the issue. Whereas the gentry of Quanzhou and Zhangzhou on the south coast, well positioned to benefit from maritime trade, agitated to keep the coast open, the gentry of Fuzhou and Xinghua prefectures further north distrusted the private wealth pouring into the southern end of the province and wanted the ban upheld. Shen supported the southern faction. Reopening trade would bring the law back into alignment with how matters actually stood.[25] The court wavered, sometimes opening the coast to supervised trade, sometimes restricting it. Even so, when the restrictions were in place and upheld, Chinese and foreign traders found ways around them. Among the telltale evidence that trade expanded after the death of Jiajing are the many Spanish American pesos ("pieces of eight" reals) that regularly turn up in archaeological excavations around the region.[26]

Despite the easing of the trade prohibition, it was still illegal for subjects of the Ming to leave the country without official permission—which few if any ever received. Exit was regarded as a rejection of loyalty to the Ming Great State. It was equally illegal for foreign traders to land on Chinese soil to do business. Accordingly, all international trade during the late Ming had to be conducted offshore: Nagasaki and Hirado on Kyushu, Manila on Luzon, Patani and Malacca on the Malaysian Peninsula, and Bantam and Jakarta (which the Dutch called Batavia) on Java, to name only a few of the more active trade entrepôts around the East and South China Seas. The one exception to this pattern of offshore trading was Macau, the small peninsula at the mouth of the Pearl River where Portuguese were permitted to come ashore in 1557 for the purpose of refitting and resupplying their ships. There they built a small port, developed relationships with local suppliers, and constructed one of the key nodes in the emerging global economy.

As trade relations proliferated through the Wanli era, public opinion shifted in its favor. Zhang Han, who was the first governor general of Guangdong and Guangxi to be appointed after the ban was partially lifted in 1567, saw for himself at first hand what trade could do. In his extended essay "On Merchants," which he may have written in the 1580s, Zhang provides a sort of economic geography of Wanli-era China. In the penultimate section of his essay he discusses the trade along the southeast coast, occasionally breaking into imaginary debate with the opponents of maritime trade.[27] "In the southeast, the foreigners gain profits from our Chinese goods, and China also gains profits from the foreigners' goods. Trading what we have for what we lack is China's intention in trade. Originally we called it tribute so that China's stature received greater honor and the foreigners became more compliant. In fact, a lot goes out and little comes back. The payment of tribute doesn't amount to one-ten-thousandth of what is transacted in trade." Zhang is exaggerating here, but his point is to emphasize that private trade is far larger and economically more important than tribute, and unstoppable. "The foreigners' minds are set on getting their profit through trade, not through imperial gifts. Even if their tribute were greater and our gifts more meager, they would still want to trade." Zhang then appeals to an old adage about "storing this wealth with the people," meaning letting the wealth that the trade generated remain in the hands of the traders rather than being monopolized by the state. He also argues that coastal security would improve if trade were permitted. "You don't understand that the foreigners will not do without their profits from China, just as we cannot do without our profits from them. Prohibit this, keep them from contact, and how can you avoid their turning to piracy?" Under the circumstances, the only reasonable policy was to accept what people were doing. "Once the maritime markets are opened," Zhang concludes, "the aggressors will cease of their own accord."

From Zhang's perspective, the tribute system was an anachronism left over from a premaritime age. The reality of the Wanli era was that private trade was operating on a scale that surpassed the tribute system. Chinese goods flowed out of the country, and foreign silver flowed in. Open the markets, Zhang argued. Let the people enjoy the benefits of foreign trade.

Prices across the South China Sea

Macau was the first hub connecting China to Europeans sailing across the Indian Ocean and trading into the South China Sea. The Spanish in their turn seized the port of Manila and made it their colony in 1571, creating a second hub for foreign trade that linked China directly to the markets and silver of Mexico and Peru. Macau and Manila along with Moon Harbor became three corners of a triangle of ports networked to each other as well as to ports further north, east, south, and west, connecting China at least potentially to every maritime nation on the globe.

Chinese mariners had already been sailing to Manila to trade for decades before the Spanish first sailed into Philippine waters in 1565. When the Spanish arrived, the Chinese were quick to determine what commodities they wished to buy. In the course of setting prices, they were careful to learn what price their exports, such as mercury, which was needed for refining silver, sold for in New Spain so as not to undersell themselves. By 1572, junks were arriving from Moon Harbor laden with a range of Chinese merchandise, and within another year or so, the trade between Moon Harbor and Manila was in full swing.[28] Predictably in the opening stage of this trade, promises were breached and payments delayed, with the result that Chinese merchants started filing complaints with the Spanish authorities in Manila. Fortunately for our purposes, a file from about 1575 survives in an archive in Seville listing the prices of goods that Chinese merchants had delivered but for which they had not been paid. The prices given in that document are not without their ambiguities, but they give a useful sense of the price regime in Manila.[29]

Table 3.2 (see appendix C, "Tables for Reference") displays the prices of twenty-five items for which Spanish buyers had failed to pay their Chinese suppliers and for which I was able to locate loosely comparable domestic prices. Sugar and buffaloes, both locally produced, were cheaper in Manila than in China.[30] Flour, pepper, buffalo calves, and footwear were on a par. Flour was locally milled, calves were locally raised, and footwear was easily produced in both economies, and so their prices were the same. If the price of pepper was loosely comparable in Manila and Zhangzhou, it is because the spice had to be shipped to both ports from the Spice Islands south of the Philippines, a trade in which neither had a comparative advantage.[31] Otherwise, Manila prices were higher than domestic. Silk and ceramics were roughly twice the price in Manila, though the price differential for the cheapest ceramics was less. Furniture was more expensive, though by a margin closer to 50 percent than 100 percent, possibly a sign that Chinese furniture makers were already operating in Manila by the mid-1570s, drawn by the boom in building and furnishing residences for the newly arrived Spaniards. As early as 1575, then, prices in the two economies were becoming aligned in ways that recognized a certain parity among basic everyday goods produced on both sides. This price alignment allowed Chinese merchants to know which value-added items to ship from China across to Manila and gave Spanish merchants access to goods at prices that were higher than in the Chinese market but not excessively so. This pricing system allowed Spanish merchants in turn to resell these goods in the Americas or Spain with the expectation of making a profit.

A second set of documents, this time from the East India Company, allows us to conduct the same exercise across several ports around the South China Sea. When the first EIC ships arrived in the South China Sea just after the turn of the seventeenth century, determining commodity prices and exchange rates was the company's first order of business in order to learn what goods to handle and where to sell them advantageously. The commanders of the company voyages were expected to compile such information and bring it back to London. One who did was John Saris. Not a gentleman by birth, Saris worked his way up from the lower rung of company employees. From 1604 to 1608 he worked in Bantam, Java, eventually rising to the position of chief factor in the trading port that the English had chosen as their base of operations in the region. The city was under the control of a Muslim ruler, but its commercial sector was dominated by Chinese, mostly from Fujian. Four years' experience in Bantam gave Saris an intimate knowledge of the terms on which trade was conducted in this economy. On his return journey to London in 1609, Saris prepared a report on commodity prices and made recommendations for trade at Bantam on Java as well as at Sukadana on Borneo. His report was part of a bid to be promoted and sent back to the South China Sea. In that bid he was successful, securing an extraordinary promotion in 1611 to the post of commander of the company's Eighth Voyage, which took him from London as far as Japan in 1613. On his return home in 1614, he compiled a second report on commodities and prices, this time focusing more on trade to Japan. Textiles predominate, a sign of their overwhelming role in driving export growth in China, followed by porcelains. Spices and aromatic woods also figure prominently among the items being traded. Combined, Saris's two reports provide a reasonably comprehensive panorama of the most heavily traded commodities in the East Asian market.[32]

Table 3.3 (see appendix C, "Tables for Reference") lists the prices of twenty items selected from Saris's schedules of prices in the three ports of Bantam, Sukadana, and Hirado on the Japanese island of Kyushu. Alongside these data I have added domestic prices, where I have been able to find them, for seventeen of the items. These data reveal first, as did the Manila prices, that domestic prices for export goods were lower than maritime-trade prices. To this pattern, however, the table contains several interesting exceptions. Three items—sandalwood, a fragrant wood; bezoar, a digestive product used for medicinal purposes; and pepper—had to be imported from Southeast Asia, which gave no advantage to domestic prices. Whereas sandalwood and bezoar were consistently lower priced outside China, pepper varied. It sold for 0.065 taels per catty in Zhangzhou, one of its point of entry into China. In Bantam a catty sold for 0.037–0.038 taels, whereas in Hirado, for as much as 0.1 taels, though Saris is careful to note that the price stayed that high only when "there come not much." A fourth exception is mercury, for which the prices in Hirado and Canton are identical. Mercury was an essential ingredient for refining silver and, in this silver-based exchange network, was in high demand. That the price of 0.4 taels per catty was the same in both locations shows that the demand for mercury was international, a regional rather than local price.[33] One further exception is iron, for which traders in Bantam and Hirado paid less than Shen Bang did in Beijing. This discrepancy may be an anomaly, as the difference may have to do with the quality of the metal these purchases record. Shen's was high-quality iron that his office purchased for making woks; the quality of the iron Saris priced may have been poorer and therefore at a lower price.

A second revelation of table 3.3 (again, in appendix C, "Tables for Reference") is that prices varied not just between China and the larger South China Sea economy but across the region. In general, Saris's prices were lowest in Sukadana, which was serviced by a smaller number of traders than the other two. They were highest in Hirado, which was the greatest distance from the rest of the South China Sea network and therefore less thoroughly integrated than the other two locations. Between these two extremes was Bantam, home to many trading communities and the region's link to the Indian Ocean. One exception is raw silk, which sold for a quarter less in Japan than it did on Borneo or Java, presumably because of the strong silk industry operating in Japan. By contrast, the price in the other two ports was identical, perhaps another piece of evidence of price integration within this international trading network.

The exercise of comparing domestic prices with Saris's prices reveals which Chinese products were profitable to sell outside the country. Commodities of lesser value such as sugar, honey, and copper sold at higher prices, though the differences between foreign and domestic prices were less than 100 percent. The category that really stands out is textiles. Gauze fetched twice the domestic price, damask up to three times, and satin from five to ten times. This was the commodity that European traders in particular wanted to buy.

If we go back a century to the prices that Portuguese merchants encountered when they first entered the Indian Ocean, it is possible to place Saris's findings on price differentials in a longer perspective. A sample of these prices appears in a report appended to a record of the "kingdoms" of India that a member of Vasco da Gama's crew, possibly Álvaro Velho, compiled after the completion of da Gama's first voyage in 1499. The report claims that its data came from an Alexandrine merchant who had traded in India for thirty years.[34] Velho examines the profitability of sixteen commodities exchanged across the Indian Ocean between the eastern end of the Mediterranean (Cairo and Alexandria) and Southeast Asia (Ayutthaya in Siam and Patani on the east coast of the Malaysian Peninsula), the central node of this trading network being Calicut on the Malabar Coast. Velho recorded prices when he had them, but he tended more often to report price differentials. For example, the differentials against prices in Alexandria for sappanwood from Tenasserim was 17 to 1, for cloves from Malacca, 20 to 1, for black benzoin from Pegu, 21 to 1, and for shellac from Patani, 31 to 1. The price differential for shipping pepper from Kadungallur, also on the Malabar Coast, to Alexandria was comparatively modest, 4 to 1, though the pepper market in Europe was so large that it ensured a profit if shipped in bulk.

When the English entered the Indian Ocean and the South China Sea a century later, these extraordinary price differentials had vanished. The growth of trade through the South China Sea and Indian Ocean economies had been prodigious, and shipping and production had increased accordingly. In contrast to Velho's price for pepper in Malacca (3.6 cruzados per quintal, equivalent to 1.2 taels per catty), Saris in Bantam was paying 3 percent of that price (0.037–0.038 taels per catty). On the other hand, since pepper in Japan could go as high as a tenth of a tael, there were still differentials to be exploited so long as pepper could be purchased and transported in large quantities.[35]

Portuguese merchants were the first Europeans to enter the maritime trading system of the Indian Ocean; they were also the first to enter the

South China Sea economy, as well as to acquire a trading post on the China coast. This location placed them in close proximity to the sources of the goods they were buying, as well to sources of information about prices that they needed to know in order to anticipate the differentials that they would seek to exploit elsewhere. For these reasons, the Portuguese were closely watched by their competitors, particularly the Spanish. Despite the temporary union of the Spanish and Portuguese crowns between 1580 and 1640, the economies of Macau and Manila were kept apart. Not a few Spaniards pressed the crown to alter this arrangement and integrate their economies. Pedro de Baeza, a merchant from Madrid, who spent a quarter of a century negotiating commercial arrangements among Malacca, Manila, and Nagasaki for Spain, was one of those advocates. As the new competitors from northern Europe entered the region, it was essential, Baeza informed the king in 1608, that Portugal and Spain (in the guise of their colonies, Macau and Manila) communicate and work together cooperatively.[36] The Portuguese in Macau should share their advantages with the Spanish in Manila, specifically access to commodities and whatever commercial intelligence they were able to collect from China, including prices.

Historian Charles Boxer speculated that Baeza was the author of an unsigned Spanish memorandum in the Seville archives listing the commodities that Portuguese merchants exported from Macau to Japan and to Goa along with their prices. The entries in this document provide a purchasing price, based either in Canton or in Macau, and then a selling price, first in Nagasaki for the eastward trade, and then in Goa for the westward trade.[37] These data indicate rates of return ranging from 40 percent to 300 percent, with a median rate of around 100 percent. Gold registered the lowest rate of return: 40 percent in Nagasaki and 80–90 percent in Goa. Silk textiles show a rate of return of 70–80 percent in both markets, with silk-cotton blends rising to 100 percent in Nagasaki, where demand was stronger than at Goa. For porcelains, the rate of return in both markets was 100–200 percent. Overall, most of the commodities that the Portuguese shipped out of Macau doubled their prices when they were carried to distant ports within the trading system of the South China Sea and Indian Ocean. They paid a modest surcharge compared to the prices of these commodities in Canton, which is where their suppliers sourced the commodities, but it was manageable given the returns. No wonder a Spanish merchant lobbied the king of Spain to take down the barrier between Macau and Manila. That barrier remained in place, however, for political reasons. More pressing, as Baeza prophesied, was the increasing presence of other European traders, which cut into the advantages that the Portuguese had enjoyed as first comers. Heightened foreign demand had the short-term effect of forcing prices up, though its long-term effect was to stimulate production, causing unit prices to fall.

Significantly, the same process of prices rising and falling appears to have been under way inside China at the same time. Matteo Ricci noted this effect in Beijing, where he was living at the same time Saris was in Bantam. The prices of foreign commodities were declining, he noticed. "Pepper, nuts [nutmeg], aloes, and other such products, which are imported from the neighboring islands of Molucca or from states bordering on China, are becoming less appreciated and are falling off in price as the supply increases." Demand was stimulating supply and driving down prices at the lower end of the market. Unaffected, though, were the prices of domestic commodities going out to luxury markets abroad. High-grade tea selling for 1 to 3 taels per catty in China fetched 10 to 12 taels in Japan, Ricci noted. A pound of rhubarb, which sold in Europe as a wonder drug "at an almost unbelievable profit," could be bought "for ten cents, which in Europe would cost six or seven times as many gold pieces."[38] From what Ricci could see, the profits to be made on the strength of domestic prices at the high end of foreign trade were not diminishing.

Prices in the Porcelain Trade

One commodity in this market that we can track closely from Dutch archives is porcelain. Porcelain had been a major Chinese export for centuries, to the extent that it carries the very name of "china" in many languages (*sīnī*, in the case of Arabic). Demand for porcelain continued and grew through the Ming, rising as trade linked Chinese production with European demand. As Jesuit missionary Francesco Sambiasi explained to his Chinese friends, "the silks and porcelains of China are what no other countries of the world have."[39] The early trade to Europe gave spices and silks pride of place in ships' cargoes. As the taste for Ming porcelain in northern Europe took off in the early years of the seventeenth century, traders responded by increasing the volume of porcelains they put in their holds.[40] When the Verenigde Oostindische Compagnie (United East India Company, or VOC), which operated out of Holland, stepped up its purchases of Chinese porcelain, Chinese merchants in Bantam worked with their Dutch counterparts to determine the styles, qualities, and quantities best suited to the European market. So quick was the response that the company soon found itself buying in an oversupplied market, such that by 1616, porcelain prices in Bantam were less than half of what they had been the year before. The VOC ended up handling pieces in the millions, and not just for the European market. Between 1608 and 1657, the company shipped over three million pieces to Europe, but that number is dwarfed by the twelve million pieces the company bought in Batavia to sell in other ports around Southeast Asia.[41]

The Dutch porcelain trade left behind abundant records, yet as Dutch historian Tys Volker so nicely put it in his seminal study of the trade, the data are "the despair of a modern statistician," complicated as they are by wholesale versus retail prices, individual versus bulk prices, prices in multiple currencies, plus elusive shipping and stevedore costs. Even so, Volker was able to determine that simple, small porcelain items such as teacups or saucers cost the VOC about 1 cent each, bowls 2 cents, with larger items running to 3 or 4 cents. More elaborate pieces such as jugs cost between 25 and 35 cents. Specialty products such as Persian-style flasks or ginger jars could cost a full tael or more. For example, the unit price for the 178 ginger jars the VOC bought in 1639 works out to just over 2 taels. The VOC paid slightly above domestic prices, though much if not all of that surcharge can be attributed to shipping costs. When Michele Ruggieri, the first Jesuit missionary to enter China, visited the porcelain manufacturing center of Jingdezhen in 1586, he found that the prices at source were lower than they were in Canton, as one would expect.[42] Shipping porcelain to Canton raised the price, and sending it abroad would have raised it again. Still, that increase was not so great that it pushed VOC prices significantly out of line with domestic prices. The per-unit price the Dutch paid for saucers, 0.7 cents, happens to be the very price Shen Bang paid for saucers in Beijing. We cannot say whether the saucers in these two purchases were comparable in quality, though teacup prices are similarly comparable. Shen's price for a teacup, 1.2 cents, falls between the VOC's cheapest bulk purchase of low-quality Quanzhou teacups at 0.8 cents and its most expensive at 2.7 cents. For the Quanzhou teacup, the better comparator might be Shen's wine cup, for which the price was half a cent. Although these comparisons are impossibly loose, they suggest that the prices Chinese exporters charged the Dutch were not hugely out of line with domestic prices. Volker in fact was surprised at how low were the prices that he found in the VOC archives.

Considered from a wider perspective, the amount of silver that the Dutch spent buying Chinese porcelains was large but not massive. Their buyers tended to go to the low end of the market in order to enhance the profit margin when prices at the other ends of the trade were good. Other Europeans shopped further up the market, however. When Florentine merchant Francesco Carletti was in Macau in 1598 on his voyage around the world, he spent the substantial sum of 20 taels to buy 650 plates and bowls of finest quality. If we compare this with Hai Rui's estimate of 1 tael for 120 pieces of everyday porcelain, Carletti paid at a rate almost four times what Hai did. This may have been a case of overcharging a naive foreign buyer, though likelier he was buying china that was substantially superior to the cheap dinnerware with which Hai furnished the residences of local officials. Carletti also bought five blue-and-white vases for 14 taels, a price that greatly exceeds all recorded VOC purchases at the time, with the exception of a few large pieces the VOC bought in Bantam in 1612 and Taiwan (where the Dutch operated a trading colony) in 1636. Perhaps what Carletti's purchase attests to is that while the Dutch made bulk purchases of everyday ware that they could sell profitably outside the South China Sea, they also traded at the high end.[43]

Not all of China's porcelain left by water. Shen Defu spent his childhood in Beijing, while his grandfather and father were posted there in the Wanli era, and recalls watching a caravan preparing for departure into the Asian interior. Shen notes that all envoys, be they Jurchens or Mongols or South Asians, were of one mind when it came to selecting what to take home on their return journeys. "They would hear of nothing but porcelain vessels," writes Shen. Some missions rolled out of Beijing with tens of carts loaded with porcelain. Shen recalls watching workers at the foreigners' hostel in the north end of the city packing one of these carts with so much porcelain that it towered nine meters above the ground. He marveled most at how the packers dusted sand, beans, and wheat kernels on every piece before they bound them together by the dozens, then sprinkled the parcels with water so that the grain and beans sprouted and swelled, packing every space so tightly that you could toss one of these bundles on the ground and nothing would break. Shen adds that "the prices they paid were ten times the regular price."[44] Even though the envoys paid through the nose, the price markup at home would be so much greater that there would be no difficulty making a good profit, exactly as the Dutch understood.

Effects of Foreign Trade on Prices

What effect did competitively priced Chinese commodities have in the markets in which they were sold? The Spanish Empire offers some intriguing evidence. Spain hoped to benefit financially from its colonies in both the Philippines and New Spain, but it was difficult to do so because of the economic relationship that developed across the Pacific between these colonies. Spain attempted to keep the Americas and the Philippines insulated from each other so as to ensure that profits from one did not leak laterally to the other rather than flow back to the empire's apex in Madrid. But the price differential across the Pacific Ocean, specifically between Chinese manufactures moving east and American silver flowing west, was greater than it was across the Atlantic. The Spanish crown wanted American silver to flow east to Spain to service Spanish consumption, not west to Manila in exchange for Asian goods that were consumed in the Americas. Not only did consumption in the Americas contribute nothing to Spain from a Hispanocentric perspective, but it also hampered Spanish exports to the Americas, thereby reducing the wealth that should have gone back to the metropole.

In 1602, the bishop of Rio de la Plata (River of Silver) Martin Ignacio de Loyola on a visit to Spain wrote a memorial to Philip II in reply to a query the king raised with him. In this memorial, Loyola argued for the importance of keeping the Indies—meaning the territories under the jurisdiction of the Council of the Indies (New Spain, which is to say, Mexico, and Peru, or South America)—"dependent on and subordinate to Spain." One way to assure that dependence was to impose it politically and religiously, though Loyola also recognized that an active commercial relationship between Spain and New Spain was essential to sustain that dependence. "If commerce should cease, then communication would cease; and should the latter cease, within a few generations there would be no Christians there." Commerce, however, had a way of leaking resources out of the empire along more profitable channels. "That which causes most injury to this commerce and communication is the diversion of the commerce . . . to other kingdoms."[45] What Loyola had particularly in mind when referring to commerce was the sale of Spanish cloth in the Americas, which the trade in Chinese cloth from Manila was undercutting. By his estimate, two million pesos of silver were leaving the Americas for the Philippines every year to pay for this cloth.[46] The effect of the trans-Pacific trade was that "all of this wealth passes into the possession of the Chinese, and is not brought to Spain, to the consequent loss of the royal duties and injury to the inhabitants of the Philippines; and the greatest loss, with the lapse of time, will be that rebounding upon the Indies [the Americas] themselves." Loyola wanted the king to limit how much any one purchaser could buy in Manila. He also called for a complete ban on silver exports to Asia to stem the price inflation it was causing in Manila, to the despair of Spanish merchants there. "The price of Chinese silks and merchandise has risen," he writes, such that "while for twenty years when only the inhabitants of the islands were permitted to trade, they were wont to gain one thousand percent, now they do not gain one hundred."[47] Cutting out buyers in the Americas and protecting buyers in the Philippines by reducing the westward flow of silver across the Pacific would restore livelihoods to the latter while bringing prices in Manila down. This correction would also cause prices on Chinese cloth in Mexico to rise to a level that would make cloth exports from Spain more competitive.

Appended to Loyola's memorial are extracts from two letters from the viceroy of New Spain, Gaspar de Zúñiga y Acevedo, dated 15 and 25 May 1602. In the first, Zúñiga registers the objection of the merchants of Lima to any restriction on their trade between New Spain and Manila. "They regard it as so necessary that, should it cease, it would mean complete destruction." The problem the merchants saw themselves facing was not "the merchandise brought to the kingdom of Peru from China" but the failure of Spain to defend their shipping from English piracy, though they also complained that import duties were too high and the time wasted on transacting customs payments too great. Another problem was that the cycle between silver leaving Lima and merchandise returning could extend to as long as three years, to which another year and a half might have to be added to clear customs before that merchandise could be sold. "Consequently," Zúñiga concluded, "this money can gain its profit only once in four years, when it could, as formerly, be thus handled twice in that time." As a result, "the merchants of Lima, who were formerly very rich and had ample credit, have become debtors," and their debt in turn was reducing the profits of merchants back in Seville. The Seville merchants raised "a cry against Chinese goods, as they imagine that to be the cause of their loss."[48] The merchants of Lima acknowledged that some of their number had moved north to Mexico in order to engage more actively in the galleon trade rather than direct their investments back to Spain, but they explained that they did so because Spain was failing to provide the infrastructure needed to protect maritime trade. They also insisted that the impact of the trans-Pacific trade was exaggerated, and that more cloth came to the Americas from Spain than from China. If they traded in Chinese rather than Spanish cloth, it was because the investment cycle of shipping from Spain was too slow, making that trade much less profitable than the China trade.

The merchants of Lima justified their trading in Chinese commodities in two more ways. The first was that Chinese cloth did not compete

with Spanish cloth, as it was "worn only by the very poor, and the ne-
groes and mulattoes [blacks] (both male and female), *sambahigos*
[children born to Chinese–native Peruvian couples], many Indians [na-
tive Peruvians], and half-breeds [children born to European-native
couples], and this in great number." Chinese cloth, presumably linen
or cotton, enabled the poor to dress themselves at low cost. In other
words, there were two markets, one for the wealthy and European, the
other for the poor, the enslaved, and the indigenous, with little overlap.
The merchants also defended their trade by observing that "the silks of
China are much used also in the churches of the Indians, which are thus
adorned and made decent, while before, because of inability to buy the silks
of Spain, the churches were very bare." The result was prosperity for all.
"As long as goods come in greater abundance, the kingdom will feel less
anxiety, and the cheaper will be the goods. The increase to the royal exchequer will be greater, since the import duties and customs increase in
proportion to the merchandise." On these grounds, the merchants re-
quested "that commerce should be opened with China, and that they
should be permitted to send one million [ducats, roughly equivalent to
Chinese taels] annually in two vessels, and that this million bring back merchandise" to the port of Callao at Lima. This merchandise could then be
sold for six million, generating a 10 percent duty for the king. The merchants
gently conceded that they could live with a limit of half a million ducats
instead of the full million if the king so insisted, which was the proposal
the viceroy supported. The core of their plea was that "the Chinese merchandise can in no way injure the commerce of Spain, while its benefit
to Peru is certain—especially to the poor and common people, of whom
there is a great number—and since it seems desirable for the adornment
of the churches of the Indians that there should be goods from China."[49]

Viceroy Zúñiga's second letter discusses problems of money supply.
Noting "a great lack of money" in Peru, he allows that this shortage
"proceeds in part from the very great sum taken out annually for China,"
yet he judges the problem to be the royal limit on coinage. If Spain
wants to redirect the flow of silver in its direction, it needs to allow Peru
to coin more silver. Zúñiga notes that the viceroy of the Philippines has
written to inform him that in Manila "goods are very dear because of the

great quantities of money that go there."[50] A note later in the file suggests that commodity prices in Manila had risen 50 percent. So while
the viceroy agreed with Loyola's dismay that "this is all in money which
goes to infidels and never returns, and thus militates against this country
[Peru] and that [Spain], and greatly weakens the commerce of both," he
advised against restricting the import of Chinese cloth in the interest of
those who could not afford Spanish cloth.

Setting aside the effect of the annual fluctuation in money supply,[51]
this discussion among Spanish administrators suggests that what Chinese merchants sold in Manila were mostly not high-end luxuries but
widely available commodities such as unadorned silk and cotton cloth,
which sold well in foreign markets because it could be priced low
enough to compete with European products. The success of that competition is exemplified by the Chinese merchandise listed in the inventories of the goods of two small shopkeepers, Juan Agudo and Domingo
de Monsalve, in a silver-mining region in northern Mexico, drawn up in
1641 after their deaths. These inventories show that Agudo stocked 3
yards of silk taffeta (*tabí de China*), and 2 yards of white *chaúl*, as well as
two garters in blue taffeta. For his part, Monsalve had on hand 5 ounces
of loose-spun silk and a pound of twisted silk.[52] The silver flowing west
to China was bringing affordable Chinese commodities east to the
Americas. Even before the Ming had ended, low-end Chinese textiles
were everyday merchandise in Mexico.

What these revelations indicate is that Chinese domestic prices were
sufficiently low compared to prices elsewhere that Ming merchandise
could be exported and traded around the world, profiting the manufac-
turers and exporters in China as well as importers and retailers elsewhere,
so long as the supply of silver was sufficient to broker the exchanges that
linked parts of the Ming economy with parts of the world.

In Favor of Trade

The merchants of Lima insisted that their trade in Chinese goods from
Manila was conducive to "the prosperity of all" and resulted in "an increase to the royal exchequer." Thomas Mun would have agreed. As a

merchant who had experienced the business benefits of international trade, he grasped the value of moving silver to markets where it received the best terms for buying goods in demand elsewhere, shipping those goods to markets where they were in demand, and profiting from the fact that the silver earned through sales exceeded the silver originally outlaid. That was the mercantilist's dream: to use the exchange of commodities as the mechanism for increasing wealth. For Mun, the increase was not only in profits for the company. He insisted that more people were able to make a living from the trade, and that the state benefited from the enhanced tax revenues.

Such arguments in favor of trade are not to be found in the Chinese documentary record of the period, not because the idea was entirely alien but because there existed no public sphere in which advocacy of trade might be voiced and given legitimacy. Chinese merchants had no public whom they could address, and they kept their views and accounts close to their chests, destroying them once the deals they recorded had been completed rather than leaving paper trails for competitors or inquisitive tax officials to follow. Mun's confidence in the confluence of private and public good required a public discourse that accepted the separation of the two and respected both. Ming officials did not have that luxury. In permitted discourse, the virtue of public good always trumped the vice of private benefit. This polarity sharpened noticeably in the years when the coastal ban was in effect. Even though the ban was partially lifted in 1567, it was revised and reimposed several times through the Wanli and Tianqi eras. When a Chinese ship sailing from Fujian to Guangdong in the mid-1620s was seized for defying the maritime ban, its cargo—which included such Southeast Asian commodities as pepper, sappanwood, and *cyclea racemosa* root (a medicinal alkaloid) valued at almost 10,000 taels—could be regarded only as contraband, that is, as private wealth being moved in defiance of the public good. Yan Junyan, the provincial judge who records the case, found that some of the cargo had spoiled, but he was unwilling to discount the cargo's value, as that would entail lightening the punishment when there were otherwise no extenuating circumstances that lessened this crime of smuggling.[53] Yet there were others in the late Ming who looked at foreign trade and argued that the underlying Confucian assumption that private wealth competed with, and thus threatened, the public good should be reconsidered in light of evidence that trade in fact contributed to the public good by increasing wealth for all.

In April 1639, five years before the unimaginable event of the dynasty's collapse, Fu Yuanchu submitted a memorial to Emperor Chongzhen requesting that the restrictions on maritime trade, reimposed in 1638, be lifted.[54] Little is known about Fu Yuanchu beyond the short biography in the "meritorious service" section of his local gazetteer. Fu hailed from an elite family from Quanzhou, the coastal port city through which most of Fujian's foreign trade from the Song to the early Ming had passed. The family possessed a significant collection of rare antiquities, signaling both wealth and high status. Fu Yuanchu also enjoyed a reputation as a scholar. He had passed the national exams in 1628 and was working in Beijing as a supervising secretary in the Ministry of Works when he offended the emperor over an appointment issue and was forced to retire in February 1638. The local gazetteer notes that "he died in office," which suggests that he was reinstated. He must have been, as otherwise he would have to have been barred from communicating with the emperor fourteen months later.[55]

In his memorial, which garnered wide notice, Fu acknowledges piracy as a problem and understands the appeal of closing the coast as a means to choke it off. But he then blunts that argument by pointing out the great advantages that would redound to the Ming if the emperor were to allow Chinese to trade with foreigners. To the west, Southeast Asians offered such luxuries as sappanwood, pepper, and ivory for which there was demand in China, while to the east, the Europeans in the Philippines had silver. Both groups wanted to trade what they had for what they were unable to produce for themselves—just as Zhang Han had argued five decades earlier to the fairly deaf ears of officials at the Wanli court. Specifically, Fu noted, the foreign buyers wanted silk textiles from Huzhou and porcelain from Jingdezhen. The Ming would lose nothing by allowing them to buy these manufactures, he argued. As the profits on both sides were great, the maritime bans succeeded in achieving nothing other than criminalizing the trade and transferring the profit to smugglers. Fu's main pitch for reopening maritime trade was that it would generate valuable customs revenue that the Ming otherwise forewent, to no advantage to itself. With the sole exception of weapons and munitions, which the Ming Code banned from being allowed out of the country, merchants from Quanzhou would best be allowed to export whatever Chinese products they could sell, thereby bringing benefits to the silk producers of Jiangnan and the potters of Jiangxi.

Toward the end of his memorial, Fu numbered off the three clear advantages of ending the ban on foreign trade. First, the Ming needed revenue to support its military defense on the northern border, which the Manchus would breach for good in 1644. Foreign trade would generate that revenue through customs receipts. Second, land along the coast was scarce and the people had only trade to save themselves from poverty. Trading with foreigners would provide them with incomes to relieve their poverty. Third, coastal military officials who might otherwise get drawn into corrupt schemes with smugglers to fund their coast guard operations would no longer be enticed into such arrangements.

Fu Yuanchu was not making these arguments in a void. He wrote his memorial in pointed opposition to those who condemned foreign trade on the grounds that it generated illegal wealth, fueled criminal activity, and opened the country to espionage. Not dissimilarly, Thomas Mun wrote his defense of EIC trade to Asia to push back against those who opposed foreign trade on the grounds that it drained England of its stock of silver, benefiting only consumers of wasteful luxuries. But there were differences. Fu was speaking against conservative bureaucrats who regarded foreign trade as the thin edge of a wedge that would weaken China, whereas Mun was agitating against an entrenched popular view that regarded the benefits of foreign trade as going solely to the corporation at the expense of the public. While Mun was fighting against public opinion, Fu was confident that he was voicing it. As he asserts at the end of his memorial, "These are not the words of one official. In reality they are the public opinion of Fujian province."[56] He closes with the assertion that he has talked to gentry and commoners alike in Quanzhou and Zhangzhou, and that everyone agrees with his view.

Mun and Fu can be connected in one further way, which is that they both mention silver in their arguments. Fu refers to it four times in his memorial. The first time is to remind the emperor that the opening of Moon Harbor to foreign trade in 1567 generated annual customs receipts of over 20,000 taels of silver, an income that had not existed previously. The second reference is to note that the price of 100 catties of Huzhou silk is 100 taels in Huzhou and 200 taels in the Philippines. The profits on Jingdezhen porcelain and Fujianese sweets and preserved fruits, he insists, are the same. Fu's third reference to silver repeats the first, this time as a loss rather than a profit. Referring specifically to the offshore trade on Taiwan between Fujianese and the Dutch, he declares that the proceeds of this illegal trade are being made at the expense of the Ming to the tune of 20,000 taels a year. His fourth reference then repeats the third: reinstate foreign trade and you reinstate 20,000 taels in the Ming military budget. They get there by different routes and assumptions, but Fu and Mun are alike in arguing that foreign trade adds to, rather than takes away from, the wealth of the country and the wealth of the people. Fu is addressing the emperor and stresses the benefits to the state: foreign trade will end the smuggling and replenish the military budget. Mun is speaking to the English public and stresses the wealth creation that international trade can bring at essentially no cost to the people. Whether either claim is accurate is another matter.

What these two voices shared was that they spoke in eras when prices were under stress. In 1621, England was in a recession. In 1639, the Ming was in far greater distress. Ten years earlier, temperatures had started to fall across the realm. Eight years into this cold phase, in 1637, drought descended over much of the land, a deadly combination that would continue all the way to 1644. Fu's home province of Fujian was struck by famine in 1639, followed by an epidemic in 1640. These calamities sowed uncertainty among officials about what policies to adopt in order to manage the financial and subsistence crises that ensued. Against those who thought that closing the borders was a sensible response, Fu Yuanchu argued that shutting the gates was the last thing the regime should do in the face of mounting difficulties. He notes widespread poverty among the people of Fujian as well as the exhaustion of the province's tax base for meeting the costs of supplying the military on the northern border. The state needed funds, the people needed income, and trade opened the path to both.

In England, Mun's view prevailed, largely because the interests of the crown were aligned with the interests of the corporation that it had authorized. In China, Fu's did not. This did not mean that either position was secure in the first case or utterly discredited in the second. As long as English coins were full blooded, the East India Company had to keep reassuring the public that taking silver abroad was not to the detriment of the national economy.[57] Mun's text was remembered, but Fu's was forgotten. It survives only because the statecraft theorist Gu Yanwu, living under the new Qing dynasty, made a point of including it in his massive compendium on state policy, *Tianxia junguo libing shu* (Strengths and weaknesses of the regions of the realm). The East India Company had the ear of the British state, whereas Gu Yanwu had the ear only of fellow scholars (his magnum opus was in fact not published until 1811). The two polities were on different courses.

A Magellanic Exchange?

Some economic historians have suggested that what drove the extraordinary prices at the end of the Ming was fluctuation in the money supply, first the influx of foreign silver starting in the Wanli era, then the strangulation of that flow in the Chongzhen era. It is certainly true that the price of grain depends on the balance between the supplies of grain and money. The price of grain can rise when demand increases, though so too when the amount of money in the economy increases. Contraction in the money supply can then affect prices in both directions—which was precisely the charge against which Thomas Mun had to defend the East India Company's policy of paying for Asian imports by taking silver abroad. At the receiving end, the argument has been proposed that the scale of new silver imports from Pacific sources, including Japan as well as the Americas, added to China's stock of money at a rate eight times greater than did domestic sources before 1600, and possibly twenty times greater through the early decades of the seventeenth century, causing prices in China to rise.[58]

This argument was an important corrective to the old notion of China as immune to what went on in the world beyond its borders. What remains in question, however, is whether the effect of the silver on money supply altered prices. As already noted, in the middle of the Wanli era, Martin Ignacio de Loyola believed that silver from Peru, flooding the market in Manila, drove up the prices that Chinese merchants charged for the goods that they brought from China. This impact led him to argue that the shipping of silver to Manila should be stopped so that the prices there might return to their earlier levels. Loyola had no opinion on whether the onward flow of American silver from Manila into China was driving up prices there, but since the global turn in Chinese history in the 1990s, some historians have insisted that it did. Do late Ming prices supply evidence for this effect?

Before answering that question, it is worth reflecting on the historiographical model driving this interpretation. That model depends on what has been called the Columbian exchange, a term coined by environmental historian Alfred Crosby to describe the exchange of biota and other materials in both directions across the Atlantic in the wake of the voyages of Genoese mariner Christopher Columbus.[59] Economist Earl Hamilton proposed in 1929 that the flow of precious metals, particularly silver, into Europe was on a scale large enough to disturb prices, destabilize the Spanish economy, and drive what he and others called sixteenth-century Europe's price revolution.[60] Political philosopher Jean Bodin was already voicing this argument as early as the 1560s, anticipating what is now called the quantity theory of money.[61] Like Crosby, Hamilton was correct in arguing that the Columbian exchange had transformative effects on both sides of the Atlantic, yet his thesis that the flow of American silver provoked the steep rise in prices in Europe has been challenged in numerous ways.

In a succinct account of these challenges in relation to money, prices, and wages in three regions of Europe during the period of the alleged price revolution, economist John Munro accepts that the flow of silver into Europe contributed to an annual inflation of between 1 percent and

1½ percent over the first 130 years of the flow of silver across the Atlantic. It did so, however, well after that inflation was already under way. The Columbian exchange was a factor, but only one of many. In addition, its impact on inflation in Spain has been shown to be less than was previously supposed. Munro has summed up his argument by reminding us that the price revolution was "essentially a monetary phenomenon, but one with national or regional variations that were the products of both local coinage debasement and, to a possibly lesser extent, the behavior of particular real forces in each local economy."[62]

The Columbian exchange has nonetheless been influential in shaping what could be called, in memory of Fernão de Magalhães, the Portuguese navigator who traversed the Pacific Ocean for the Spanish monarchy in 1520–21, the Magellanic exchange. Silver was shipped in volume from Acapulco to Manila and exchanged there for Chinese products, after which it was moved from Manila to Zhangzhou and from there penetrated the Ming economy. If the Columbian exchange is proving to be less reliable than historians used to think, a Magellanic exchange stands on even weaker legs. First of all, at the level of fact rather than theory, more silver was reaching China from Asian sources, particularly Japan, than from the Americas. Second, while the Manila galleons plying the trade route across the Pacific did create an important channel for the exchange of Chinese commodities and American silver, the bullion that reached Ming China was as likely to arrive from the other direction, after it had crossed the Atlantic to Europe and then was transshipped by agencies such as the East India Company to Asia. Finally, the Ming economy was large enough—roughly equivalent to the entire economy of Europe—to absorb the arriving silver into its systems of commercial exchange rather than be destabilized by its addition.

This is not to say that the arrival of silver in excess of 100,000 kilograms per year had no effect whatsoever on the Ming economy. But that effect has yet to be empirically demonstrated. The only sector in which I have been able to detect the effect of an abundance of silver driving up prices, noted toward the close of the preceding chapter, is the luxury market. Participant-observers of the Wanli era attributed the spectacular rise of painting prices to the surge in the number of new buyers entering a limited market. The question this raises is not where they came from, but how this growing group had the abundance of silver to participate in the luxury market. I have found no indisputable evidence that higher art prices are a consequence of increased silver entering the country from abroad. Finding such evidence will require new research on the late Ming art market. The strongest case I can make at the moment is the logical one that the "unworthy" buyers so disparaged by the cultural elite, who felt threatened by their decreasing ability to monopolize the possession of status objects, were men whose "new" wealth came from the expanding commercial economy, who chose to store that wealth not simply in the relatively inert asset of land but in physical silver, and who were more than content to exchange that economic capital for the cultural capital of highly valued items—which in any case could readily be converted back into cash when circumstances required. That this surge in high-end luxury consumption should have excited critical commentary in the decades immediately after the rise in legal maritime trade in 1567 argues in favor of at least the plausibility of this hypothesis.

But art prices are not the subject of this book; grain prices are. To understand what pushed grain prices up late in the dynasty, we now turn from the luxury economy to the famine economy. The argument that imported silver was the principal factor forcing up prices at the end of the Ming is far weaker than the argument that I am about to present in the next two chapters, which is that global climate, not global trade, drove Ming grain prices to crippling levels.

4

The Famine Price of Grain

AT LONG LAST, we arrive at the core phenomenon around which I have written this book: the prices to which grain rose in times of dearth. Famine prices do not appear in Ming documents with any consistency before 1450, yet from that date forward, they begin to pile up over the following two centuries, making up the longest series of prices in China before the eighteenth century. Unlike ordinary prices, famine prices are exceptional. They express the particular, sometimes even peculiar, circumstances of when and where they were recorded, but their common denominator, from which flows their value as consistent data, is that they registered the gap between what people expected to pay for grain and what they would have to pay to buy it, to a degree that Ming observers all understood. If we can use these deviant prices to write a price history of the Ming, it is because, once braided into a long ribbon of data, they display more clearly than any other proxy from Ming documents the pressures under which people found themselves, baffled and distressed, during severe climate downturns between the mid-fifteenth and mid-seventeenth century. In addition, these prices give us nearly direct access to the limits of what was possible, to invoke Fernand Braudel once again. When grain rose to a price beyond what most people could afford, these prices marked the hard boundary between the possible and the impossible, on the wrong side of which no one wished to be. What disrupted that boundary was not the supply of money. It was the erosion of the natural conditions for agricultural production during what we know as critical phases of the Little Ice Age.

In this period of history, the price of grain was the most reliable barometer of agricultural prosperity and human survival—and also of political stability. When in 1420 Emperor Yongle opened his interview with emissaries from the Timurid ruler Shahrukh Mirza by asking about conditions in Persia, he started by inquiring whether grain was dear or cheap in their country. When the envoys assured him that it was cheap, Yongle graciously declared this to be evidence that Shahrukh enjoyed Heaven's favor.[1] Low grain prices meant abundance, which was a sure sign of Heaven's approval—a theological matter on which Yongle as a usurper was sensitive.[2] At the time of the audience, Ming grain prices were also low. Other than a major famine in 1406 following several years of excessive rainfall, and the flood years of 1415 and 1416, temperatures stayed in the normal range, and abundant precipitation ensured good harvests.[3] Agricultural prosperity gave Yongle the means to undertake many expensive projects, including rebuilding the Grand Canal, relocating the Ming capital from Nanjing, where his father had reigned, to Beijing, and sending a series of diplomatic armadas to the Indian Ocean. The dynasty's run of luck would start to fade a decade and a half after his death in 1424. But during his reign, grain was cheap. He was Heaven's choice.

The people of the Ming believed, as the emperor did, that the world was in good order when the price of grain remained stable and fair. When Chen Qide rhapsodized over the price of grain during "bountiful harvests and prosperity," he was giving voice to the shared understanding that this was how the world should be. Everyone accepted that prices fluctuated seasonally, falling to their lowest right after the harvest and rising to their dearest during what was called "the time when green and yellow do not meet," which is to say in early summer, when the old (yellow) grain stocks were depleted and the new (green) grain had yet to be harvested.[4] But as soon as the next harvest came in, they were confident, the price would return to normal. That was the expectation. Starting in the middle of the fifteenth century, however, that expectation was undermined every few decades by crop failures that shattered price certainty, at least in the short term. Not until the end of the dynasty was it shattered in the long.

The Price of Grain

In the opening year of the dynasty, 1368, the founder ordered that a list of valuations be compiled to guide magistrates when assessing penalties for theft, as penalties were based on the value of the thing stolen. The list includes five grain prices. The most expensive is rice at 3⅛ silver cents (31¼ coppers) per peck. Next comes wheat at 2½ cents (25 coppers). Third in price is millet at 2¼ cents (22½ coppers), which was also the price of beans. The cheapest grains on the list are sorghum at 1½ cents (15 coppers) per peck and barley at 1¼ cents (12½ coppers), though neither was regarded by contemporaries as food fit for human consumption.[5] No other evidence confirms these prices, but nor does any deny them. The need of the new regime to establish its legitimacy in the eye of the public argues in favor of regarding these valuations as correct and fair, which is why I repeat these figures as reasonable approximations of the opening prices of grain in the Ming.

The rarity of references to grain prices in Ming sources means that we do not have such a clear set of prices for any subsequent year, though random references from later in the dynasty suggest that these prices continued more or less to prevail. For example, in his report of prices in Canton in 1608 referred to in the preceding chapter, the Madrid merchant Pedro de Baeza recorded that the price of rice there ranged from 3½ to 4 cents.[6] Remarkably, the lower end of that range is only ¼ cent above the valuation made two and a half centuries earlier. Chen Qide could similarly assert that during his childhood in the early Wanli era "the price of a peck of grain was only 3 or 4 cents." The upper end of this range, 4 cents, is probably the better approximation of the price of rice in the sixteenth century, but both observers testify that it was more like an upper limit. A report to Wanli's grandfather in 1566, for example, notes that when grain is cheap, its price stays below 4 cents a peck, though when it is expensive it can rise to 6 cents or more.[7] Even so, a report to Emperor Wanli in the 1580s begins with the observation that "the price of rice in Jiangnan does not go above 30 cents" per hectoliter, or 3 cents per peck. The official who submitted the report, Zhao Yongxian, concedes that prices elsewhere may be a little higher but insists that on the

Yangzi delta "the price of rice is extremely cheap."[8] A few years later, in 1588, Zhao reported to the Grand Secretariat that "after the harvest, the price of rice at its highest does not exceed 50 or 60 cents" a hectoliter, or 5 to 6 cents a peck. This might look like a significant leap in the normal price of rice, but it is not, for Zhao wrote this report just as the price of rice in Jiangnan was beginning to come down after the first great Wanli-era famine the previous year when, as he notes, rice had been driven up to the exceptional price of 1.6 taels per hectoliter, or 16 cents per peck.[9] Chen Qide was not wrong when he recalled 3 to 4 cents as the proper range for the price of rice.

Grain prices did not always stay within that range, however. The poet Ren Yuanxiang could recall that "in the old days of the Wanli era, a hectoliter of grain was 40 to 50 cents [4 to 5 cents per peck] and everything was cheap."[10] Other writers of the period cite 5 cents per peck as the standard price of rice. As early as the 1530s, for instance, Tang Shunzhi observed that the standard price of rice in the Suzhou region was 5 cents per peck.[11] Looking back from the 1660s after the Ming had collapsed, Lu Wenheng recalled that a hectoliter of rice in his youth in Suzhou cost "only 50 or 60 cents," or 5 to 6 cents per peck.[12] Another Suzhou writer, Liu Benpei, looking back from the opening decades of the Qing, recalled that a peck of rice was 5 cents toward the end of the Wanli era, and that only in 1622 did the price start to rise.[13] A contemporary report of a famine in Jiading county outside Shanghai in 1624 corroborates 5 cents as the normal price of rice "in a good year."[14] These testimonies have referential value, though in many cases they are retrospective prices that the writers cite in order to evoke life as it was before the huge surges in the price of grain late in the dynasty.

On the basis of these scattered references, we can say that the price of rice started at just over 3 cents a peck in the early Ming and more or less retained that price for over a century. Late in the sixteenth century, 3 cents tends to be cited not as the price of rice so much as the lower end of a range that went up to 4 cents. Occasionally 5 cents is named this early, though that price is not recorded with any consistency until the 1620s, by which time, as we shall note later in this chapter, the people of the Ming had gone through two waves of subsistence crises during the Wanli era that had knocked the foundation out from under the assumption that the price of grain could be expected to remain as it always had been.[15]

Famine Grain Prices in Local Gazetteers

The main source of the famine grain prices used in the book is a quasi-official genre of local history and geography known as gazetteers. Published as substantial books, often in multiple volumes, gazetteers chronicled the history, statistics, public affairs, and prosopography of a local area. Most commonly that area was a county, the lowest reporting level of Ming administration with a population in the tens of thousands. Gazetteers were updated and often substantially revised in new editions, which were published ideally on a sixty-year cycle. The gazetteer was not a genre mandated to report prices, to the dismay of some commentators, but it did aspire to provide a complete, permanent record of local events that included abnormalities and disasters as well as achievements and glories.[16] Accordingly, when local prices deviated exceptionally from normal, the genre treated these prices as worthy of inclusion in the historical record. As a result of this orientation, the only references to the price of grain were preserved as evidence of just how far out of line grain prices had diverged from what they should be.

Famine was not the only occasion that might inspire gazetteer compilers to record the price of grain. So too were superabundant harvests that drove prices to extreme lows. Through the first two centuries of the Ming, the bumper price of rice is fairly consistently reported as 2 cents per peck, which is roughly two-thirds to a half of the regular price. The bumper price crept up over time, such that by around 1570, 3 cents emerges with the same frequency as 2 cents as a standard bumper price in places where the normal price might be 4 cents or higher.[17] The bumper price for millet was less than for rice and, with only a few exceptions, does not rise above 2 cents.[18] When bumper prices are recorded in copper, as they often are, they start early in the Ming at 7–10 coppers, rising later in the sixteenth century to 20–30 coppers. The rise in copper prices is steeper than it is for bumper prices in silver, possibly reflecting a difference between retail prices (in copper) and wholesale prices (in silver). Bumper prices for millet start at the same level, rise to 10 coppers in the 1490s, and then follow rice upward in the sixteenth century though with a certain lag, not reaching 30 coppers until the seventeenth century.[19]

Bumper prices were a boon for consumers, but could be a disaster for producers who relied on the income from selling part of their harvest. By the same logic reversed, famine prices were a disaster for consumers but could be a windfall for producers. Not a single reference in Ming sources to extreme prices notes the differential effect of price distortion between consumers and producers. It would appear, accordingly, that famine and bumper prices were recorded from the perspective of buyers, not sellers. Bumper prices simply made grain cheaper to buy, to the delight of buyers, and famine prices made it more expensive to buy. We are given no indication of how grain producers experienced price distortions. That noted, the two perspectives would have met in the most extreme famines, when climate conditions were so severe that no grain could be harvested and grain farmers found themselves in the same position as buyers. At that point, which is the point at which a famine became worthy of recording, producers and consumers faced the same intolerable price.

The prospect of famine loomed over every premodern agrarian economy. If the people of the Ming had an advantage over other peoples, it was in living under an administration that actively championed the public good and maintained a system of grain storage explicitly to adjust the price of grain in times of dearth. The dynastic founder was keenly aware of the vulnerability of the poor to famine grain prices. As his biography in the dynastic history reports, "1344 was a year of drought, locusts, major famines, and pandemic. That year, when he was sixteen, his father, mother, and [three] elder brothers died one after the other, and he was too poor to bury them."[20] When Zhu Yuanzhang founded his regime twenty-four years later, in 1368, he was determined that the price of grain should be low and stable. Price stability, he believed, characterized the golden eras of China's past. "When the Tang controlled the realm, the weather was seasonable and the harvests plentiful," he told a Hanlin academician in the patent of appointment he issued to the man. "A peck of grain cost 3 coppers, and there was a sufficiency for every family and person. Once having heard of this, Our heart leapt in hope of being adequate to the task."[21] He aspired that his dynasty should align in this way with the great dynasties of the past.

In reality, Zhu could not hope to drive down the price of a peck of grain to 3 coppers. The most he could hope for was that, relative to the prices of other commodities, the price of grain would stay within the buying power of ordinary people. Associating Tang prosperity with good weather expressed his belief that good weather must follow good government, as Heaven blesses the ruler who devotes himself to the well-being of his subjects. But he also understood that his role should include building and stocking granaries to offset bad weather when it came. In the third year of his reign, he ordered that every county construct four permanent granaries.[22] Zhu might not be able to force the price down to 3 coppers, a move that would have impoverished farmers whose livelihood depended on selling their grain at the current market rate. But he could task his administration with the goal of ensuring that grain should never be unaffordable. Accordingly, one of the tasks laid on the shoulders of the local magistrate, as the prefectural gazetteer of Guangzhou (Canton) notes, was to "keep the prices of things stable."[23]

Magistrates dreaded price rises just as much as ordinary people did. Magistrates would not starve, but if the people had no food, the desperate might take matters into their own hands by causing disturbances and seizing the grain stocks of the wealthy, which would flag the magistrate as unable to maintain local order and possibly end his career. The fear of disorder was real. As a central official notes in a memorial on famine relief early in the sixteenth century, anxiety that order might collapse during a famine rises at the same rate as does the price of grain.[24] The local magistrate's simplest mechanism to restabilize prices when famine struck was to release grain from the county granary, assuming it had any. One compiler of a county gazetteer in Shandong province argued a century after the Ming that grain prices would have not gone to the unaffordable level they did at the end of the dynasty had stocks been available to release to the famished.[25] This was true in theory, although in practice most local granaries were empty before the Ming fell. Another option was to turn to local wealthy families and appeal to them to donate their stocks of grain or sell them at reduced prices. The prefect of Songjiang did exactly this during a famine in 1630, eliciting the response he hoped for. "Men of virtue banded together and compiled a register in which they subscribed the grain they would make available to a volume of not less than ten thousand hectoliters in order to level the price. As a result, this year the price of grain did not soar."[26]

This combination of responses was usually sufficient to bring prices down in all but the worst crises.[27] When these responses were absent or failed, the consequences could range from flight to starvation, and beyond these, to banditry and even cannibalism, which was the worst effect that people of the Ming could and did imagine. When this happened, extreme price data of all sorts come into view. One sort was the price for which a parent could sell a child, which fell dramatically. Denominated in grain, the price of a boy could range from 2 pecks of grain down to as little as 3 liters.[28] Another sort of extreme was the price at which grain became so expensive that the only thing to eat were people.[29] The price trigger of cannibalism varied by place, time, and circumstances. In one county in Henan province in 1588, cannibalism started when grain reached a price of 200 coppers per peck. In the same county in 1640, the starving did not resort to cannibalism until the price of millet had reached 1,500 coppers per peck.[30] Elsewhere in the same province a year earlier, the price trigger during a locust infestation was 1.4 taels of silver.[31] A county gazetteer in Shandong province records the trigger for cannibalism in 1640 as 2 taels.[32]

Local gazetteers were where these deplorable facts were registered. For one example, at the end of the entry regarding a "strange famine" in 1639, the compiler of the gazetteer of Gushi county in Henan province lists these prices in half-size characters (the Chinese equivalent of the footnote): "The price of a peck of grain," which in Gushi meant millet, "was 3,500 coppers, of a peck of wheat, 2,500 coppers, and of a peck of barley, 2,000 coppers."[33] For a second example, the compiler of the Shanxi provincial gazetteer records the effect of a famine two years later by giving prices in silver, again in half-size characters: "A peck of wheat went from 8 or 9 tenths of a tael to 1.2 taels, and even 1.5 or 1.6 taels." This gazetteer also—and this price information is extremely rare—records elevated prices for vegetable oil and pork, both at 8–9 cents per picul.

Disastrous prices were exceptional facts that gazetteer compilers recorded to commemorate trauma rather than to track economic conditions, and indeed they had to be truly exceptional to be considered worthy of recording. Regarded as inauspicious, compilers usually sequestered these prices among such Heavenly, Earthly, human, and animal abnormalities as the births of human triplets and two-headed calves in a section variously entitled *Xiangyi*, "Auspicious Signs and Anomalies," or *Zaixiang*, "Disasters and Auspicious Signs," and usually buried toward the back of the book.[34] A strength of the gazetteer genre is the practice of dating events, auspicious or inauspicious, to the year, sometimes to the month and day. And when fuller notices of the event existed, they could be appended to the entry, usually in half-size characters, which is how Chen Qide's famine memoirs of 1641 and 1642 managed to reach modern historians.

A particular value of price reports in gazetteers is that they quantify the intensity of famine. Almost never does a compiler record the other obvious quantifier, the number of people who died, probably because this is not a statistic that was easily compiled. Price is therefore the only metric that registered the scale of a disaster. To be clear, it was not standard practice to include a famine price with every famine report. The compiler might mention that the price "leapt" or "surged upward," but he was under no obligation to record the price to which it rose. Providing a price, if he had a record of one, was a device by which the compiler could signal the severity of the disaster. Although famines prices are infrequently recorded, they provide values that can be aligned with and tested against other such values and subjected to some measure of statistical analysis. These extreme prices often come with a brief annotation as to the reason the price rose to the level it did, "flood" or "drought" being most common, though occasionally the annotation reads simply, "for no reason."[35]

To illustrate the kinds of price information a local gazetteer might preserve, consider the reports of famine in the 1633 gazetteer of

Haicheng county—the county created around the port town of Moon Harbor in southern Fujian in 1567 just before Emperor Jiajing died and Emperor Longqing agreed to reopen the port to maritime trade. The 1633 county gazetteer reports that Haicheng experienced a sudden shortage of rice in the spring of 1615. The price shot up to a dizzying 20 cents per peck in early June. Magistrate Tao Rong responded by disbursing loans of relief grain from the county granary. This grain fed the people of his county for a week, long enough to give grain barges coming south from Zhejiang time to arrive and replenish the local market. Tao was fortunate that he did not have to rely on locally stored grain in the long term, as his granary held only enough rice to bridge the dearth for a week until confidence in the local market was restored and grain dealers were able to bring commercial grain into the locality. Once the commercial grain arrived in Haicheng, "the price of grain returned to normal." Tao was fortunate in having some grain on hand, but major famine relief in this period depended more on the capacity of the market to piggyback on state relief efforts, as well as on the government's willingness to incentivize private merchants to provide what it could not by such means as adding a legal surcharge to the wholesale price.[36] When famine hit Haicheng again in 1630, local supplies were not sufficient to ease the surge in demand, and supplies did not arrive from elsewhere. According to the gazetteer, "a peck of grain commanded a price of 20 cents, the bodies of the starved littered the road, and people were reduced to eating the leaves off trees. It took a year for the price to return to normal."[37]

From the scattered records that I have extracted over the course of skimming some three thousand gazetteers of the Ming, Qing, and Republican periods come the 777 famine grain price reports on which this chapter is based. What these price records report, though, is not entirely obvious. Were these prices what anyone actually paid, or were they used to mark a level above what most people could not afford to buy grain? No gazetteer compiler I have read comments on the nature of famine prices or how he derived them. They simply appear as local historical facts. Despite such hesitations, these prices have a specificity and consistency that allow for aggregating them into a data set for Ming history unlike any other.[38]

The Distribution of Famine Prices

The famine grain prices that are reported in gazetteers can be tagged in four ways: by type of grain, by currency, by place, and by year. The types of grain represented are mainly millet and rice, with some wheat. Very occasionally one encounters famine prices for the cheaper grains of buckwheat, barley, and oats, though so rarely that I have excluded them from the data set. Of the grains that are tracked in this chapter, rice predominates. There are roughly seven rice prices for every two millet prices, and nine for every wheat price. Which grain went to a famine price in a gazetteer entry is often not specified, the entry simply using the generic term, *mi* (kernel). Turning to other sections of the gazetteer sometimes reveals what the likeliest grain for the county was; and when that was not the case, I fell back on the convention of dividing Chinese agriculture at the 800 millimeter isohyet (the line connecting points having the same amount of annual rainfall), which roughly follows the southern boundary of Shandong province: millet to the north and rice to the south.[39] Finally, while extreme prices were recorded in both silver and copper, prices in silver predominated by a rate of seven to four.

Famine price reports in local gazetteers come more from the eastern and the northern areas of the Ming. Gazetteers from South Zhili (anchored around Nanjing on the Yangzi delta) yield 20 percent of the reports, followed by North Zhili (anchored around Beijing on the North China Plain) with 15 percent, Zhejiang with 11 percent, and Henan with 9 percent. Least represented are the southwestern provinces: Yunnan yielded only seven reports, Guangxi six, Guizhou four, and Sichuan three. Between these two extremes fall the price reports in gazetteers from the northwest, the west, and the southeast. This distribution does limit the claims that can be made about climate having a uniform impact across China during the Ming, but not, I think, to the point of diminishing the overall argument of this book.

A test of whether this distribution is an artifact of the sources rather than a reflection of regional climate variation is to compare the provincial distribution of price reports with the provincial distribution of extant gazetteers.[40] The comparison reveals that South and North Zhili account for 14 percent and 10 percent of gazetteers respectively, which shows that these provinces' reports of famine grain prices outperform their share of the genre by roughly 50 percent. So too do the reports from Zhejiang and Henan, where famine price reports outweigh their share of gazetteers by roughly 25 percent. The geographical distribution of gazetteers thus modestly distorts the geographical distribution of the data. To some extent, this distortion reflects the distribution of population. The fit is good at both ends of the scale. South Zhili had the largest share of Ming population, possibly around 17 percent, and Guangxi, Yunnan, and Guizhou were the three least populated provinces. On the other hand, there is little correlation among the provinces between these extremes. North Zhili, with only 7 percent of Ming population, had the second highest number of famine price reports, presumably because of the presence of the capital there and the concern for dynastic stability.[41] Of course, the most salient factor shaping the distribution of price data is the incidence of famine, which may have been higher in the more densely populated eastern provinces and the more environmentally fragile northern provinces, though that research remains to be done.

More critical for the analysis this book undertakes than the geographical distribution of famine grain prices is their temporal distribution. Space may be the stage on which these prices were set, but time organizes the rhythm by which they unfolded. The chronology begins with one price report in 1373, the sole report from the fourteenth century, and ends with eleven reports in 1647—three years beyond the end of the dynasty but nonetheless part of the price disturbance at the end of the Ming. The other 765 extreme price reports lie scattered among 142 of the 244 years from 1403 to 1646 inclusive. The details of the chronology can be tedious, but the evidence is in the details, and the denser they are, the clearer the story we can tell about the travails of living through the Ming.

The first report, in 1373, is followed by single reports in 1403, 1404, and 1428. These four price reports make up the sum total of famine price reports through the first seven decades of the Ming. Reports are few in the early years of the dynasty in part because the practice of keeping records and publishing local gazetteers took decades to become well established, though the absence of references to price disturbances in other sources suggests that the early period of the Ming was a time of stable prices. Not until the early 1440s do we get the first (small) cluster of four prices from the years 1440–42. The first big cluster comes the following decade, between 1450 and 1456, when there are no fewer than twenty reports of extreme grain prices, far exceeding the number from the preceding eight decades. The recording of disaster prices at this rate of three per year will be wildly dwarfed by reports in the sixteenth and seventeenth centuries, but in the context of the time, when such reports were only beginning to be recorded, it marks off this era, which coincides precisely with the reign of Emperor Jingtai, as a period of significant grain shortage. Though no Ming author makes this observation, I would suggest that the crisis of prices in the 1450s was responsible for initiating the practice of keeping a local record of aberrant grain prices.

After the Jingtai era, the incidence of famine price reports is irregular. There is a cluster of five reports for the years 1464–67, another five reports for the years 1471–72, and four for 1478–79. The next surge comes in the 1480s. Over the nine years from 1481 to 1489, I found thirty-nine famine price reports. There is a one-year hiatus, then another six reports for the years 1491 and 1492. Extreme price reports appear erratically through the decades after 1492, with small clusters in the years 1507–10 and again between 1512 and 1516. The next great surge is spread over a quarter of a century: thirty-four reports between 1520 and 1526, twenty-four in the years 1528–32, twenty between 1534 and 1541, and finally an unprecedented forty-three reports in the three years 1544, 1545, and 1546. There is a single report in 1550, another cluster of seventeen reports between 1552 and 1554, and then after a three-year break, twenty reports for the five years from 1558 to 1562. A scattering of price reports can be found over the next twenty-two years, followed by a strong surge to a new peak of reports through the second half of the 1580s, with seventy-four reports over six years. This wave is followed by five smaller waves across the three decades from 1596 to 1625.

Emperor Chongzhen ascended the throne in 1627, the first of an unbroken procession of twenty-one years in which not a year goes by without a large number of extreme grain price reports, for a total of 316 down to 1644, the year in which Chongzhen committed suicide and the dynasty collapsed. With this block of data I have included another eighteen reports from the three years following the fall of the Ming, which lie outside the last Ming reign but serve to identify the tailing off the Chongzhen crisis. Within that stretch of time from 1627 to 1647, the peak period is the four years from 1639 to 1642 when the annual average of famine price reports soars to fifty-six. In no other period in the Ming, and probably in no period at any other time in Chinese history, had disaster struck on this scale. It is no wonder that Chen Qide twice put his pen to paper. The data for this chronology are summarized in table 4.1 (in this chapter). Not all 777 price reports in the data set are counted there, only the 712 reports from the years that were continuous with other years of price reports. I have clustered them in this way to highlight the major periods of agricultural crisis and price forcing.

Now for the prices that were reported. The normal copper price of rice through much of the Ming was 25–30 coppers per peck. The most commonly cited copper famine price through to the late years of the sixteenth century is 100 coppers—three to four times the normal price. This was something like the "standard" famine grain price through that period. A few reports of a price of 200 coppers can be found in the fifteenth century. (Very occasionally a price of 1,000 coppers is recorded, though these reports may be errors resulting from reporting a price per hectoliter as the price per peck.) By the 1540s, 200 coppers emerges as a second "standard" famine price alongside 100 coppers. After the turn of the seventeenth century, 1,000 coppers is the price most frequently reported. By the late 1630s, multiples of a thousand appear regularly. In 1640, the year for which I have found the greatest number of famine prices in copper, thirty-six, most fall within a range between 1,000 to 2,000 coppers, though some rise as high as 10,000. That level continues among the twenty-four prices in 1641, most of which range from 1,000 and 5,000 coppers. The number of prices recorded for 1642 falls, as does the price, which is mostly in the range of 600 to 700. In 1643, however, famine prices are back up in the thousands of coppers. I have found only one copper price from 1643, which is 4,000 coppers. The following year, 1644, the year the Ming fell, extreme grain prices are back up in the

TABLE 4.1. Clusters of Years in Which Famine Grain Prices Were Recorded, 1440–1647

Reign era	Reign years	Cold	Dry	Officially recorded famine years	Continuous years of famine prices	Number of price reports
Zhengtong	1436–1449	1439–1440	1437–1449	1438–1445	1440–1442	4
Jingtai	1450–1456	1450–1455	1450–1452	1450–1457	1450–1456	20
Chenghua	1465–1487	1481–1483			1481–1489	39
Hongzhi	1488–1505		1482–1503		1491–1492	6
Zhengde	1506–1521	1504–1509		1507–1514	1507–1510	7
Jiajing	1522–1566	1523		1524	1520–1526	34
		1529		1529–1531	1528–1532	24
		1543		1538	1534–1541	20
		1545	1544–1546	1545	1544–1546	43
				1553	1552–1554	17
			1558–1562		1558–1562	20
Wanli	1573–1620	1587	1585–1589	1587–1588	1585–1590	74
		1595–1598	1598–1601	1598–1601	1596–1602	23
		1605	1609		1606–1609	14
		1616–1620	1614–1619	1615–1617	1614–1620	23
Tianqi	1621–1627				1622–1625	10
Chongzhen	1628–1644	1629–1643	1637–1643	1632–1641	1627–1647	334
Total appearing in this table					94	712
Overall total					144	777

A

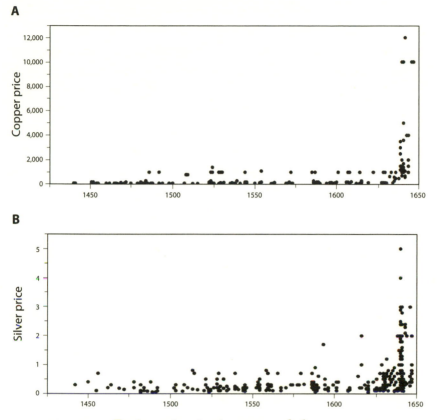

FIGURE 4.1. Famine grain prices in copper and silver, 1440–1647:
(a) Bivariate fit of copper price by year; (b) Bivariate fit of silver price by year.

thousands of coppers, with an average of a little over 2,000 coppers per peck. Two centuries of these prices are presented in the first graph in figure 4.1.

Switching now to silver prices, the normal price of a peck of rice through much of the Ming was 3–4 cents, rising to 4–5 cents during and after the Wanli era. Unlike copper prices, however, no "standard" famine grain price in silver emerges. Until the 1540s, the price fluctuates between 10 and 20 cents, averaging about 15 cents, three times the normal price. Reports of 30 cents begin to appear in 1545, after which the price ranges from 16 up to 30 cents into the 1580s. Famine prices push up to higher levels during the late 1580s. Thereafter they fall somewhat, not returning to 30 cents until the 1610s. In the 1620s, the famine price slips down to under 15 cents. By 1630, however, it has risen to 50 cents (half a tael) or higher. The year 1639 is the first in which prices reach and even exceed a full tael per peck. In some areas the famine price of rice stays below half a tael, yet in others it exceeds 2. In 1640, the number of price reports rises to its highest, eighty-three, spread across a range from 0.5 to 4 taels. The forty-four prices from 1641 occupy a slightly lower range, from 0.3 to 3 taels. The fifteen prices from 1642 keep to the low end of that range of 0.3 taels, never exceeding 0.5 taels. Chen Qide's two price reports for rice in the early 1640s align flawlessly with this profile: over 2 taels in 1640 and 0.4 taels in 1641. He gives no price from 1642, for there was no rice in Tongxiang to buy. Of the seven prices from 1643 (five rice, one millet, and one wheat), the upper end of the range rises to 2.4 taels, for an average of 1.6 taels. I have found only two silver prices for 1644, both rice, averaging 1.2 taels. The second graph in figure 4.1 displays the movement of disaster rice and millet prices in silver between the 1440s and the 1640s.

Now that we have the data, it is time for the interpretation.

Heaven, Climate, and Famine

Grain has a price when it is exchanged for money. The three main factors affecting that exchange are the supply of grain that is available for sale, the number of people seeking to buy it, and the supply of money available for that purpose. An increase in demand can push prices up, as can a surge in money supply, though not so much in the short term. What drives grain prices up in the short term is a fall in supply. As noted in the previous chapter, there is no overwhelming evidence that the influx of silver was having much of an impact on grain prices in the late Ming. The focus of the analysis in this chapter will be on the impact of climate fluctuation on grain supply. In taking this approach, I echo historian Christopher Dyer, who concluded from his research on the history of the mediaeval English economy that "grain prices reflect a number of influences, including the supply of money, but the quality of the harvest provided the main cause of the sudden surge in prices."[42] To state the

obvious, the factors that dominate the quality of the harvest, temperature and precipitation, are what the climate delivers. This is my approach
in what follows: to align the major fluctuations in Ming grain prices with
changes in climate. And so it is to Ming climate history that we now
briefly turn.

China's dynastic rulers were keenly attentive to environmental disruption, although they would not have phrased their concern in this
way. What concerned them was the idea that famine was a manifesta-
tion of Heaven's displeasure rather than the effect of our abstraction of
climate change. Since only the emperor had the authority to communicate with Heaven and interpret its actions, official historiography was
tasked with recording the disturbances that Heaven visited upon the
emperor and his people so as to maintain a record of what Heaven did
as evidence of the human failings that provoked its interventions. These
records for each dynasty were then summarized in the official history
produced in the wake of every dynasty's collapse by the successor regime. The dynastic histories have been the first stop for climate history
since Chinese historians began to take up the subject in the 1930s.[43]
Favoring a narrative of rise and fall, these histories begin with the rise
of the dynamic founder supported by a cadre of followers and end with
the fog of disorder that descends on the realm because of indecisive
rulers and incompetent officials. The Ming does not require much trim-
ming to fit this narrative structure, emerging as it did under the leadership of a vigorous founder and collapsing in a perfect storm of imperial
misjudgment, bureaucratic infighting, and popular rebellion.

Accompanying every dynastic story is an account of the perturba-
tions of Heaven, Earth, and humankind that mark its rise and fall. The
authors of the official history of the Yuan dynasty, compiled in the early
years of the Ming, give climate—again to use our term—a role in this
narrative when they preface the section on natural calamities with the
observation that "humans in their relationships to Heaven and Earth
make a three-cornered world: each has its place in the arising of disaster
and good fortune."[44] When all is well, Heaven ensures that the conditions
pertaining to the atmosphere are benign; when all is in chaos, Heaven
unleashes storms and celestial terrors to strike fear into the hearts of

those whom it wishes to put on notice. This is why climate anomalies were worthy of recording, sufficiently so to be given their own chapters in the dynastic history. It is Heaven, judging a dynasty's failures, that causes temperatures to fluctuate, rain to fall, hurricanes to blow, lightning to strike, and dragons to wreak havoc. Heaven is not alone in the Confucian cosmology as the bringer of catastrophe. Earth does the same, on Heaven's cue. It produces seismic quakes, unleashes floods, and spawns plagues of insects. Humans do the rest by murdering parents, assassinating rulers, and waging wars. All are equally signs and agents of a disturbed cosmic order that invites a new regime to take over.

The compilers of the dynastic histories recorded these disasters as proxies not of climate but of the health of the realm and the fortunes of the ruling house.[45] The events they preserved for posterity—floods, cold, snow, frost, hail, thunder, locust infestations, dragon sightings, epidemics, celestial drumming, waterlogging, changes in river courses, black miasmas, lack of snow, urban fires, celestial fires, excessive rainfall, rat infestations, ice storms, droughts, white miasmas, tornadoes, windstorms, earthquakes, landslides, and famines—cover what climate historians look for, weighing heavily as they do on the two key issues for agriculture, warmth and water. The individual entries are telegraphic, consisting only of date, location, and incident, and rarely exceed one line of text, but the work the editors did to categorize these events and list them chronologically makes them a convenient data set from which to begin reconstructing larger environmental trends. Some entries feel more random and undersubstantiated than others, yet they all had to clear an unstated but well-understood bar of severity to justify being included in the dynastic history. The editorial process thus imposed a certain measure of consistency on the data, enhancing their value for our analyses.

In earlier research, I made extensive use of the disaster records in the dynastic histories, in addition to disaster chronologies in fourteen provincial and prefectural gazetteers, to reconstruct a pattern of climate change and environmental stress over the four centuries of the Yuan and Ming periods.[46] Figure 4.2 displays the years for which I found deviations in temperature and precipitation in the documentary sources,

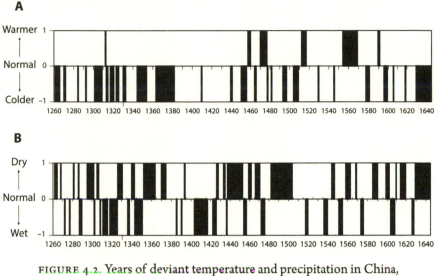

FIGURE 4.2. Years of deviant temperature and precipitation in China, 1260–1644: (a) Warmer and colder years; (b) Drier and wetter years.

from the founding of the Yuan Great State in 1260 to the collapse of the Ming in 1644. (Some of this information has been inserted in the third to fifth columns of table 4.1 to help readers correlate the clusters of extreme grain prices with climate disturbances.) That research indicated that China through this period experienced abnormal climate conditions of the sort that historians of Europe associate with the Little Ice Age.[47] Through much of the Yuan and Ming, China was a cold country. Between 1250 and 1450, according to the documentary record, China registered only a single year (1312) as abnormally warm. In contrast, sixty of the other two hundred years were abnormally cold.

Temperatures plummeted in the first year of the rule of Emperor Jingtai in 1450 and remained abnormally low through to the year in which he was deposed in 1456. The timing coincides strikingly with physical climate proxies gathered elsewhere. Based on the evidence of tree-ring data, astronomer John Eddy proposed in 1976 that a decline in solar activity drove the earth into a colder phase between 1450 and 1550. He named this phase the Spörer Minimum in honor of Gustav Spörer, a nineteenth-century astronomer who studied sunspots. China's experience of colder temperatures through the latter half of the fifteenth century indicates that the Spörer Minimum applies to China as much as to the rest of the Northern Hemisphere. These colder temperatures may have been exacerbated in the 1450s by extensive volcanic eruptions in the southwestern zone of the Pacific Ocean, from New Zealand to the Bismarck Archipelago and Java up through Luzon and Japan.[48] The local effect of blocked solar radiation was so intense that, in the summer of 1454, canals on the Yangzi delta were choked with ice. The following winter, harbors froze, canal boats could not move, and animals perished under snow that accumulated to a depth of a meter.[49] Cold weather continued to plague China intermittently until the mid-1550s, when the Ming experienced fifteen unbroken years of warm weather (1554–68), the longest spell of abnormally warm weather in three centuries. After that, the climate shifted back to cold. From 1569 to 1644, there were only three years when temperatures rose above normal. The cold became so severe in 1577, for instance, that lakes on the Yangzi delta froze and winds blew ice into mounds ten meters high.

Two decades later, in the mid-Wanli era, Jesuit missionary Matteo Ricci in his account of traveling south from Beijing on the Grand Canal in the winter of 1597–98 was moved to comment on the severity of winter temperatures. "Once winter sets in," he observed, "all the rivers in northern China are frozen over so hard that navigation on them is impossible, and a wagon may pass over them." Ricci was puzzled as to "why the great rivers and lakes of north China should freeze thick in the winter." He speculated that it might have something to do with "their proximity to the snow-capped mountains of Tartary," as though the cold were a permanent regional condition that required explanation, not an anomaly—the doing of Earth rather than the working of Heaven, in Chinese cosmological terms.[50] Had he not left Europe in 1577, just prior to the start of the worldwide fall in Northern Hemisphere temperatures, Ricci might have realized that China's cold weather did not require a local explanation. The winters were just as cold, and just as unexpectedly so, in both locations: the Thames froze in England as did the Grand Canal in China, though not necessarily in the same years.[51] Through the last half century of the Ming, only one year of abnormal warmth (1602)

was reported as against twenty-three years of abnormal cold. Temperatures fell even further in 1629, remaining abnormally low through to
1644 and beyond.[52]

Precipitation through the Ming displays a bit less volatility than temperature. Overall, however, rainfall tended to be low. Drought dominates
the record, though not to the complete exclusion of wet years.[53] Starting
in the mid-fourteenth century, the climate was normal to dry until the
sodden years of the Yongle era early in the fifteenth century.[54] Through
the first half of the dynasty, forty-six years were dry, twenty-eight wet.
Not until 1504 did precipitation return to normal. There followed pluvials (periods of abundant precipitation) during 1517–19 and 1536–39. In
1544, however, China was hit with three years of severe drought. A prefectural gazetteer in Zhejiang reports that "the lakes dried up completely
and became reddened earth." The price of grain skyrocketed, such that,
in the language of the gazetteer entry, anyone lucky enough to buy even
just a liter of rice risked being murdered as he carried it home.[55] In the
century from 1544 to the end of the Ming, China experienced thirty-
one years of severe drought compared to fourteen years of wet weather.
The worst droughts were during 1585–89 and 1614–19. In 1615, the fields
were so parched, according to an entry in the dynastic history, that the
landscape looked burnt.[56] Drought had not yet reached its nadir, however. That was what happened through the dynasty's last seven years,
when China experienced its most extended period of extreme drought
in centuries, possibly even in the millennium.[57]

Documentary proxies thus reveal that China between the 1450s and
the 1640s was consistently colder and intermittently drier than it had
previously been. The combination of cold and dry proved deadly for
grain agriculture. Human factors, including the disruption of commercial circulation, can precipitate famines, but China through these two
centuries was largely a society where labor was abundant, grain was
stored, and markets functioned, and where large-scale violence was
limited to the rebellions and invasion that brought down the dynasty.
For disaster to strike on the scale and pace that famine grain prices suggest, it had to have occurred in the context of extreme climate events.

Six Sloughs

To reduce the noise in the record of climate disturbance, and as well bring to the fore the human impact of these climate changes, I have identified six multiple-year periods during the Ming when severe temperature or precipitation abnormalities or both coincided with documentary reports of environmental crisis, famine, and social distress. For these periods of harsh conditions, I use the term "slough" (which rhymes with "bough" in British English, "through" in American, and either in Canadian). To each slough I have assigned the title of the reign era during which it occurred, distinguishing the two sloughs of the Wanli era as Wanli I and Wanli II.

As almost no famine grain prices from the Yongle Slough (1403–6) are on record, we will begin the narrative with the next, the Jingtai Slough (1450–56). As previously noted, only as of 1450, the first year of the Jingtai reign, do records of famine prices begin to appear in local gazetteers with any regularity. Emperor Jingtai was given the throne to replace his half brother, Emperor Zhengtong, who was taken hostage by a Mongol army in 1449 and who was later returned to Beijing to live under virtual house arrest until his faction seized back the throne from Jingtai in 1456, the year after the great famine of 1455. The Jingtai emperor could not have chosen a worse half decade to rule, which is why the faction supporting him was easily pushed aside. Climate does not on its own dethrone emperors, but it would be irresponsible to treat the coup against him as though it had nothing to do with the environmental stresses of his years on the throne.

The weather continued cold and dry through the rest of the fifteenth century, the Spörer Minimum. These conditions increased in severity in the early 1480s and the late 1510s, though without precipitating a crisis equal to Jingtai. The next major crisis did not arrive until the drought of 1544. The stress of that drought was compounded the following year by a wave of cold, producing a stretch of environmental crisis that I have tagged the Jiajing Slough (1544–45). Although severe drought returned in the late 1550s, temperatures remained at or above normal, and the crisis eased.

Cold weather struck again in the 1580s, fifteen years into the reign of Emperor Wanli. The era coincided with the global deepening of the Little Ice Age, leading in the second decade of the Wanli era to the most severe famine of the sixteenth century, a period I have designated the first Wanli Slough (1586–89).[58] The crisis was massive, starting in the north in 1586 and worsening through the cold year of 1587 when it spread to South Zhili and Zhejiang, inflicting famine, floods, locusts, and epidemics. By 1588, much of China was overwhelmed. Officials in distant Guangxi province reported that "people are eating people and the corpses of the famished are scattered about unburied. In the cities and countryside are scenes that even a truly talented painter, were he here, would be unable to paint."[59] That Europe also experienced a severe famine crisis during these years reminds us that this slough was global in scale.[60] Temperatures warmed in 1589, though not until the following year did the drought lift. As we know from Wanli's conversation with Lady Zheng, when famine threatened Henan province in the spring of 1594, it was feared that the disaster would return, though sustained cold did not set in until the following year and drought until three years after that, and not to a degree of severity that generated a crisis as extreme as Wanli I.

The weather turned cold and dry again in the 1610s, tipping China toward the Second Wanli Slough (1615–20). The year 1614 was marked by drought in some regions and flooding in others.[61] By the autumn of 1615, petitions for relief were pouring in to the court from all quarters. On 25 November, two grand secretaries explained to the Wanli emperor that "although the situation differs in each place, all [reports] tell of localities gripped by disaster, the people in flight, brigands roaming at will, the corpses of the famished littering the roads. There is not one report that does not plea to receive the favor of your imperial grace." Another illustrated famine memorial reached the emperor in 1616 from Shandong province, estimating that over nine hundred thousand people were on the brink of starvation. The memorialist declared that local relief had run out and that civil order had collapsed, though unlike the album sent in during the anticipated Henan famine in 1594, this device was not enough to provoke a personal response from Wanli.[62] The famine spread to the Yangzi valley later that year, Guangdong province the next, and made its way into the northwest and the southwest the year after that.

A clear-eyed observer of the Second Wanli Slough is Wu Yingji. Wu lived in the Yangzi River town of Guichi some two hundred kilometers upriver from Nanjing. Eight times between 1615, the year to which I date its onset, and 1644, the year Beijing fell to the Manchus, he took passage downriver to the southern capital to sit the provincial examination. He failed every time. Fortunately for historians, he kept a memoir of his sojourns in Nanjing, in which he mentions the famine prices he encountered in the city. Wu frames these disasters within what he knows of earlier price distortions. "Since the beginning of the dynasty," he was told, "rice in Nanjing became expensive only once or twice in the Jiajing and Wanli eras, reaching as much as 2 taels" per hectoliter. These would be references to the Jiajing and Wanli I Sloughs. Wu notes that the price went to 1.6 taels in 1588 during Wanli I, "though that lasted for only a month or two." Wu writes that he acquired this information from the celebrated Nanjing essayist Gu Qiyuan, whom he met in the southern capital during Wanli II. Gu told him that 1.6 taels was the rate at which the government granary sold grain, though despite that intervention, the market price went to 2 taels. "The elders have said that in the past two hundred years, the price of grain in the southern capital never rose to this level." What Gu told Wu was true: this was the highest price Nanjing residents had ever seen. Fortunately, it did not last, and Wanli II came to an end.

Emperor Wanli died in 1620 just as the wave of cold, dry years subsided. Emperor Tianqi muddled his way through a seven-year reign that is regarded as one of the worst-governed periods of the dynasty, though perversely enough, the weather returned to something close to normal during this time. When Tianqi died in 1627, he was succeeded by his sixteen-year-old brother, enthroned as Emperor Chongzhen just as the climate began to deteriorate. From this year forward, every price pinnacle that anyone could remember was then dwarfed by the next to arrive. Back in Wanli II, Gu Qiyuan and Wu Yingji could still take comfort in the idea that the price of grain would return to normal. When Wu Yingji wrote his reminiscences from the perspective of the disasters of the Chongzhen era, though, the prices of Wanli II seemed mild compared to what came later. "Through 1640, 1641, and 1642," he wrote, "the price went up without cease, reaching 3.6 taels. In the surrounding prefectures and counties, though, it was even worse." Looking back from the Chongzhen Slough (1638–44), Wu could regard the Wanli II as a time of innocence when people suffered from high prices but could still look forward to those prices dropping to normal levels in due course. Back then, he observes, "city residents had no idea what buckwheat or barley even were. Now these grains are selling at 5,000 cash per hectoliter. In Shandong and Henan, however, the price of millet is 10,000 cash per peck"—per peck, not hectoliter—making it twelve times higher than in the capital. However terrible prices in Nanjing were, Wu judged that living in the capital "by comparison was like living in paradise."[63] How expectations can shift.

Had Wu Yingji succeeded in his final pilgrimage downriver to write the civil service examination in 1644, he would have found himself credentialed to work for a regime that no longer existed. He had no way of knowing that, as he was witnessing the collapse of the Ming price regime, he was also witnessing the collapse of the Ming political regime. Beijing fell to peasant rebels in April 1644, then to an invading Manchu army. Nanjing held out for another year, but eventually it too had to surrender to the Qing armies that rolled south and consolidated the Manchu military occupation of China.

Price Forcing

The price of grain in any one year is dependent on the environmental conditions under which grain is grown, conditions that in a preindustrial economy are most strongly determined by climate. In their research on the impact of climate on prices in Germany during the Little Ice Age, Walter Bauernfeind and Ulrich Woitek found that the amplitude with which grain prices fluctuated under climate pressure was twice that of the prices of other goods. This is not surprising, given the centrality of grain to diet and to the economy as a whole. They also concluded, not surprisingly, that climate had greater impact on prices during periods when climate deteriorated than under conditions of mild deviation.[64]

Rather than use climate to derive prices in China, I propose to work in the other direction and use famine prices to detect climate change, then to analyze the consequences for economic and social life. For prices are not merely economic indicators of the factors affecting grain production and consumption; they are social facts that document the relationships people have not just with the environment but with each other as they buy and sell, and in the course of doing so provide or fall short of providing for themselves and their families. The vulnerability of social relationships to extreme grain prices during disturbed climate periods is often heightened by the compounding effects of other disturbances that climate can induce, from floods to locust infestations to pandemics. I should note finally that the correlation between prices as effect and climate as cause is anything but constant, not least because climate tends to be regional whereas prices tend to be local. On the other hand, while the grain trade may be available to alleviate price fluctuations, severe climate disruption can destroy a harvest so extensively that the resources of a region cannot compensate for the shortfalls of a locality.[65]

The effect of climate on prices is most clearly visible in the final half century of the Ming, from the two Wanli Sloughs to the Chongzhen Slough. Wanli I was sudden and severe, driving prices to levels the people of the Ming had not previously experienced, yet it was short enough that the grain and wealth that had accumulated over the benign decades of the second half of the Jiajing era were sufficient to support a fairly quick recovery. Additionally, it had the salutary effect of putting the entire administration on crisis watch, which explains why Emperor Wanli and Lady Zheng became involved in famine relief in 1594. Wanli II has left fewer price proxies (only twenty) compared to Wanli I (seventy-eight). The sharpest increases were for silver prices for both rice and millet, most conspicuously in 1616. Other indicators, though, suggest that the slough lasted at least until 1619 and more likely into 1620, given the nine silver price reports that I was able to find for that year.

Possibly the strongest impact of the Second Wanli Slough was not on famine grain prices in China but on the rise of the Jurchens north of the Great Wall. The great Jurchen unifier Nurhaci was submitting tribute to the Ming throne as late as 1615, but the drought and cold persuaded him to change tactics, escalating his competition with the Ming, especially for the grain grown in Liaodong in the northeast. In the cold, dry year of 1618, he launched an attack in eastern Liaodong that gave him complete control of the region. The Ming countercampaign the following spring collapsed at the battle of Sarhu on 14 April 1619. Still, another quarter century had to pass before the Manchus seized their opportunity to invade and conquer China.

The next slough came quickly. There were short bouts of colder weather late in the Tianqi era. In January 1627, the last year of his reign, snow fell on the Yangzi delta to a depth of close to two meters. "Bamboo and trees snapped," the Songjiang gazetteer reports, "and birds and animals died in great numbers." Drought started in the following year, the first of Chongzhen's reign.[66] Temperatures across the country plummeted the following year. The first serious crop failure in Songjiang prefecture was in 1632, when "the price of grain soared" and people starved. Conditions eased slightly through 1634–35, but in 1637 the temperature dropped again. That January, the weather turned "so extremely cold that the Huangpu River and Mao Lake both froze." These conditions ushered in the Chongzhen Slough (1638–44), the worst seven years of the dynasty, or indeed, of the millennium.[67] North China was hit first. Not until 1640 did the great downturn reach the Yangzi delta, first with a massive infestation of locusts followed by a great drought. The early grain crop wilted in the fields for lack of water. In June farmers planted a bean crop, but massive downpours in July washed that away. Farmers then planted a third crop, which died because not a drop of rain fell for the rest of the year. That winter, the famine that Chen Qide describes began. The entry for 1641 in the Songjiang prefectural gazetteer records sandstorms, locusts, severe drought, and price rises. "The price of rice and millet leapt, and the corpses of the famished lay in the streets." Locusts returned in the spring of 1642. Through the summer months of 1643, not a drop of rain fell.[68] Our sample of disaster grain prices confirms the intensity. Of the 777 prices in the set, 32 percent fall within the seven years from 1638 and 1644 (see figure 4.3). Even though the weather returned to something close to normal in 1644, rebels pushed out of the northwest by famine captured Beijing and drove the emperor to suicide, on the heels

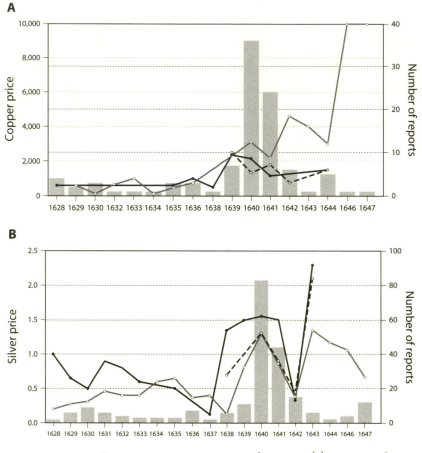

FIGURE 4.3. Famine grain price reports and prices in (a) copper and (b) silver for rice, millet, and wheat, 1628–47.

of which the Manchus crossed the Great Wall and absorbed China into the Qing Great State.[69] Political instability rode on the back of the uncertainty of the weather, and unstable prices continued to be reported for three years after the fall of the Ming dynasty.

Just as the reports of famine prices during the Chongzhen Slough are not evenly distributed across China, so the prices vary regionally. If there is a distinction to be drawn between the higher and lower silver prices in this decade, it is between the harsher north and the agriculturally more prosperous Yangzi River valley and further south. In 1638, silver prices in the south rose to no higher than 18 cents, whereas those in the north started at 70 cents. The difference between north and south diminished in 1639, with the ceiling for southern prices going up to 20 cents and for northern prices coming down to 50. In 1640, famine prices in the north and south divide at 50 cents: higher in the north, less in the south. The 50-cent ceiling held for the south in 1641–42, but prices in the north started at 80 cents and rose from there. The difference between north and south widens further in 1643–44: southern famine prices stay below 40 cents, whereas 2 taels becomes the lowest famine price of grain reported in the north.

Prices in copper do not display the same clear geographical division, perhaps because copper prices more directly express dearth in a particular place rather than across a region, whereas silver prices express something more like a regional wholesale rate. To summarize the course of copper prices in the north through the Chongzhen Slough, I offer this chronology from the gazetteer of Neiqiu county in a remote corner at the south end of the North China Plain:[70]

1628–1629: Drought for two years; 1 peck of grain cost 160 coppers.

1635: Famine; 1 peck of grain cost 200 coppers.

1638: No wheat harvest in the summer; no millet harvest in the autumn; 1 peck of grain cost 500 coppers.

1640: Spring drought; of a hundred homes, every one was empty; people dug up roots and stripped the trees of almost all their bark; no wheat harvest in the summer; 1 peck of grain cost 720 coppers; no rain in the eighth or ninth month [September–October]; no wheat planted.

1641: Still no rain as of the sixth month [July]; a peck of grain cost 1,200 coppers; in the market grain was sold by the cupful. The people had nowhere to flee. When a young man and young woman met, it was not to copulate but to cannibalize. Mothers ate their children and children, their mothers. Husbands ate their wives and wives, their husbands. There wasn't a day when people did not starve to death, die of the epidemic, or suffer execution. Alas, that human nature could be annihilated to this extent. On the 29th of the 6th month [5 August] it started to rain. In the 7th month [August] buckwheat and wheat were planted, leaping in value to 2,600 coppers, and later 3,600 coppers. By the 10th month [November] people were eating weeds; a child could be sold for 1 peck of grain. . . . From ancient times nothing like this has ever happened.

As Geoffrey Parker in his study of the global crisis of the seventeenth-century world has observed, the Northern Hemisphere of eastern Eurasia through the early 1640s was subject to largely the same climate extremes, and to the same social and political disruptions, as was western Eurasia.[71] The Chongzhen Slough in Neiqiu county and beyond was part of this global phenomenon, though it may have been more severe than elsewhere, or than at any time before this, because it coincided with a burst of volcanic activity along the Pacific Ocean's so-called ring of fire from Chile in the southeast to Japan, the Philippines, and Java in the southwest, in exactly those years. China has almost no volcanoes, but it sits downwind from a string of volcanoes through the islands off its eastern coast. Dozens of volcanoes erupted through the Chongzhen Slough, spewing massive volumes of matter into the atmosphere. The wave of explosions started on the Japanese islands of Kyushu and Izu in 1637, followed in 1638 by the eruption of Mount Usu on Hokkaido and Mount Raung on Java. There were no volcanic eruptions in Asia in 1639, only in Italy and the Cape Verde Islands, but they resumed in Asia in 1640: Komage-taku on Hokkaido, Mount Parker on Mindanao, and Mount Awu on Sangihe Island in Sulawesi, as well as Mauna Loa on Hawaii. The following year, 1641, saw major eruptions on Java, Luzon, and Mindanao (Mount Parker again), and in Japan, in addition to an eruption on Deception Island in the Antarctic. These were followed in 1642 by the eruption of Sakura-jima on Kyushu and Miyake-jima on Izu, in 1643 by major eruptions on the volcanic islands of Karkar (formerly Dampier Island) and Manam (formerly Hansa Island) off the coast of Papua New Guinea, and in 1644 by Mount Asama on Honshu.[72] Not only did this wave of volcanic activity block solar energy from reaching the earth's surface, but it also triggered a reversal of the annual monsoon in a climate disturbance known as the El Niño Southern Oscillation for two years running in 1638 and 1639, and then again in 1641 and 1642. The monsoon rains that normally fell on China and southeast Asia were swept east and dumped along the far side of the Pacific, resulting in massive flooding in the Americas and massive drought in China.

By depriving farmers of the warmth and precipitation their fields needed to produce grain, these climate distortions forced the people of the Ming beyond the limits of the possible. Crop failure meant that grain that had been held over from previous years, if there was any, commanded impossibly high prices. Drought not only parched the fields; it also emptied the canals on which the barges of grain merchants and the state might otherwise have floated food to distressed areas. Climate did not just force prices to rise but altered the entire system through which grain was produced and marketed.

The history of famine grain prices in the Ming thus turns out to provide a set of dated proxies not just for climate change but for the interaction between climate variation and the human condition. In that fatal final decade of the Ming period, extreme prices were part of an environmental crisis that precipitated economic collapse, domestic rebellion, and foreign invasion. Famine grain prices do not explain the fall of the Ming, but narrating the final great crisis of the Chongzhen era in the absence of climate would be a tale told by an idiot, full of sound and fury, signifying nothing, to quote Shakespeare. More than that, the prices point to the cause. What had China in its grip was not moral failure but climate failure. The scale on which the climate deteriorated made the fall of the dynasty as irreversible as any morality tale could imagine. Chen Qide credited Heaven with the disaster; we credit climate change. Whatever analytical framework we apply, however, nature must take the credit. Perhaps other leaders or other administrations could have devised means to ameliorate the crisis, but that is almost a pointless counterfactual when grain could not grow in the fields. The political regime collapsed, as did the price regime, and neither would return.

5

The Chongzhen Price Surge

IN THE OPENING CHAPTER, I quipped that the reader could regard this book as an extended footnote to Chen Qide's record of the disasters of 1640–42. I was not being entirely facetious, for this is more or less what I have done. If Chen needs footnoting, it is because he did not write for us. We who read this book are not Ming Chinese. We live scattered across a world that Chen made no attempt to imagine. Also, we inhabit a globe four centuries past the Ming. At the time he wrote, climate change was beyond understanding. Even Chinese readers are foreigners to the world Chen inhabited four centuries in the past. What Chen wrote, he wrote for his contemporaries. He meant his text to stand as a memorial of a time of suffering so great that it should not be allowed to fade from the collective memory of Tongxiang natives into the next generation. If it faded, it would be to their peril, Chen was sure.

What Chen Qide wrote would have made immediate sense to his intended readers. They inhabited the same price regime, knew what things should cost or at least what they had cost in earlier times, and had a visceral appreciation of how appalling were the prices that Chen records. If his account makes intuitive sense to us, it is because we understand the larger climate context in which these events occurred. Between how the people of the Ming understood their experience in relation to the wrath of Heaven and how we contextualize it in relation to the Little Ice Age, the gap is considerable. The preceding chapters served to establish the Chinese context for what things cost, how households strove to manage to meet their costs, and what happened when climate downturns destroyed harvests and forced prices to rise. My purpose in this concluding chapter is to set Chen Qide's experience in relation not just to the fall of the Ming dynasty but to the longer course of China's history. What do we as historians do with this record of prices, beyond noting their role in the destabilization of the political regime? How do we model what happened? Was the Chongzhen Slough simply a seven-year aberration of purely short-term significance? Or does it have a place in analyzing not just the circumstances surrounding the main political event but the longer trend of change that Chinese experienced through and beyond the seventeenth century?

One model that comes to mind is the "price revolution" that Europe experienced in the sixteenth and seventeenth centuries, which we reviewed in chapter 3. This model proposed that the silver that flowed in great volume across the Atlantic increased the money supply at a rate the European economy could not absorb, thus driving up prices. These higher prices altered the terms on which goods were exchanged, labor was paid, and capital formed, causing widespread distress. The "El Dorado," or Realm of Gold, that Spaniards thought they had found in the New World ended up counterintuitively producing not unlimited wealth but economic chaos. Historians do not dispute that prices rose in seventeenth-century Europe. What they have called into question is the monetary argument at the heart of the thesis, which is that the importation of silver from Spanish colonial possessions in the Americas was what caused that upward shift. To defer to the critical judgment of economic historian John Munro once again, inflation in early modern Europe was due not solely to money supply but to many factors, including coinage, population growth, and state financial practices, each of which had a different effect depending on the local economy in which these factors operated. It was not predetermined that the influx of silver should drive up prices in all economies, nor cause all prices to move at the same rate.[1] Under the scrutiny of what actually happened on the ground in Europe, the high-level theory of the price revolution falls apart.

David Fischer has proposed a lower-level theory of price revolution, which he calls the price wave. His model has had some influence among historians who, like me, are not economists but are keenly aware of the need to assess the financial environment that shaped the lives of their historical subjects. According to Fischer's model, prices move upward significantly through a long-durational process he calls a price wave, which unfolds in five stages. The wave starts when prices begin to rise slowly during a period of prolonged prosperity and demographic growth, to which he credits such good things as "material progress, cultural confidence, and optimism for the future." In the second stage, in Fischer's terms, prices break through "the boundaries of the previous equilibrium," usually in the context of wars of ambition or dynastic struggle, producing price instability and political disorder. In the third stage, states and individuals respond to rising prices by seeking to expand the money supply, a proposition that reverses the usual cause-and-effect of an increasing money supply forcing up prices. As the expansion of money supply runs into limits, as it inevitably must, the fourth stage brings price chaos, market instability, falling real wages, and fiscal crisis. The wave then breaks, ushering in the fifth stage, which can include regime collapse, population decline, and falling prices, leading eventually to a new equilibrium.[2]

Munro among others has faulted the wave model for ignoring the particularities of how and where price inflation takes off, and also for giving demographic change a leading role to the exclusion of monetary variables, the impact of international trade, and the role of financial institutions. Nor is Munro persuaded that the periods before and after an inflationary wave are best characterized as equilibriums.[3] Despite these limitations, I find it worth asking whether the wave theory could be said to have a certain formal similarity with the rise and fall of prices leading up to and beyond the Chongzhen Slough. My purpose in asking is not to endorse the wave model or to bring China into line with the price-revolution paradigm. It is simply to reorganize what we have learned thus far and consider the shape of the curve of prices through the seventeenth century. Some of the questions to be considered are whether the surge of grain prices in the 1630s and 1640s was preceded by a modest upward movement in prices; whether the rapid price rises in the Chongzhen era constituted a permanent breakthrough in the price regime; whether these price rises induced a third-stage expansion in the money supply; whether these stages led inexorably to price chaos and political collapse; and finally whether the fall of prices after the Chongzhen era stabilized into a new equilibrium.

Just stating these questions in plain language reveals that not all of Fischer's stages align neatly with the record of Chinese prices. The most serious lack of alignment is at his third stage, which posits that rapidly rising prices would have led to efforts to expand money supply. The problems here are that bullion started flowing into China before a rapid rise in prices, and that no evidence suggests that grain prices during the Chongzhen Slough affected the movement of bullion. The regime indeed collapsed, as the fourth stage of the Fischer model predicts it should. But in the absence of the destabilizing expansion in money supply of the third stage, the logic that the one was the consequence of the other dissolves. If we remove these two pieces of the model, there may still be something to learn by looking for what the first two stages of the wave model predict, and then by skipping to the fifth stage and considering whether prices reached a post-Ming equilibrium.

Starting with the first stage of the Fischer model, does the Ming economy demonstrate a long phase of a moderate price rise? To test this proposition, I will produce estimates of long-term rates of inflation over the first two centuries of the dynasty using a series of four sets of prices between 1368 and 1590. For the second, I will examine a set of prices from the middle years of the Wanli era around the turn of the seventeenth century to ask whether we can detect a more rapid rate of price change preparatory to the price chaos of the Chongzhen era. These exercises should help to determine whether a process of internal price inflation was already under way prior to the peak of the wave, and whether the Chongzhen surge should be regarded as an outcome of that process.

Moving past the collapse of the Ming regime, we can then ask whether the model helps to capture what happened after the leap of grain prices in the Chongzhen era. Was the Chongzhen surge a short-term aberration? Did it lead, in wave fashion, to a new price regime based on Chongzhen prices? Was there a long-term restabilization? Chen Qide certainly hoped that the price chaos he was experiencing was temporary, though he did worry that after the disturbance ended, the survivors would simply congratulate themselves on having got through the crisis and "then turn their attention entirely to acquisition and enjoyment" as though nothing had happened and everything was simply back to the way it had been, leaving themselves unprepared to deal with the next crisis when it came. To place the events of the Chongzhen era in a longer context, we will consult the price data collected from the 1630s to the 1690s by yet another writer on the Yangzi delta, Ye Mengzhu.

The point of this exercise is not to treat prices as independent variables of economic history, but to use them to evaluate the impact of climate on economy and society. To tell the final great crisis of the Chongzhen era only in terms of climate would be a simplification, though to tell it without climate would be to elide the essential context. The value of placing climate change close to the center of the story is that it helps to focus attention on the impact that such change had on the lives of ordinary people. Scientists working from physical proxies have done much to track the climate fluctuations we have followed from documentary proxies, yet no climate simulation can determine exactly the moment when, in Ye Mengzhu's phrase at the end of his note on the price catastrophe of the 1640s, "the people were utterly exhausted."[4] This is why the climate history of the Ming cannot be told in the absence of price history—nor its price history in the absence of climate history. Each lends local precision to the other, and together they alert us to the ultimate dependence of preindustrial society on climate, which is to say, its vulnerability to variations in the amount of solar energy reaching the earth's surface, on which farmers depend to grow the food societies need to reproduce themselves. Turning to rates of inflation can help to decipher the significance of price changes in the contexts of long-term climate change as well as short-term climate disturbance, as the one manifests itself through the other.

Long-Term Price Change through the Ming Period

To test the hypothesis that climate changes forced up prices, we need to consider whether other inflationary or deflationary factors were at work in the background. To detect rates of inflation, I propose to make three comparisons of prices across time using four data sets, from 1368, 1451, 1562, and 1590.

The first set, and the baseline for all three comparisons, is the earliest recorded price list of the Ming: the list of monetary values that guided magistrates when they needed to assess the value of property in order to determine severity of punishment, which was graded according to the value of what was stolen or destroyed. Issued in the year the dynasty was founded and preserved in the dynastic statutes, this list of prices assigned to 264 items is not a record derived from actual transactions. It is a compilation of values assigned by the state. Despite their legislated character, we can be reasonably confident that these prices reflected the prices people actually paid. The list was published as part of the new regime's bid for legitimacy as well as legal authority. It was burdened, therefore, by the popular expectation that the prices set would be fair approximations of what things actually cost. An additional challenge with using this source is that the prices therein are denominated in paper scrip, a Mongol financial instrument that quickly lost its value over the course of the founding emperor's reign. However, a "string" of what was known as Treasure Scrip was exchanged by fiat—and it was a strong one—for 1,000 copper coins in 1368, the equivalent of one tael of silver.

The second set of data is another government list, issued in 1451, of the duties to be levied on commodities entering the city of Beijing. This list specifies duties on 226 items. The standard duty on commercial goods was 3 percent ad valorum. The Beijing tax effectively doubled the duty to 6 percent in a bid by the administration of the newly enthroned Emperor Jingtai to get a grip on the regime's perilous finances after its dramatic losses to the Mongols in 1449. We can calculate the prices for 1451 by multiplying the 6 percent duties by a factor of 16.7. Like the 1368 list, the prices on this list are denominated in paper scrip, which by 1451 was basically useless as money and purely an accounting device. The fact that both lists use the same unit of currency does make the task of comparing prices from the two documents relatively straightforward, assuming that the official nominal value of that unit has not changed.[5] Of the 226 entries on the 1451 list, I have identified seventeen items that are sufficiently similar to items on the 1368 list to warrant an exercise in comparison. These prices are displayed in table 5.1 (see appendix C, "Tables for Reference").

The exercise yields some simple but useful observations. First, between 1368 and 1451, prices moved but not all in the same direction. Some rose, some fell. Second, the rate at which prices of goods of a certain type moved varied hugely by item. In the case of metals, for example, the price of lead fell by 44 percent whereas the price of copper rose by 180 percent. This difference may reflect the difficulty of sourcing metals but could as well point to an increase in the cost of energy. Lead has a low melting point and is cheaper to process than a harder metal such as copper, which requires greater energy to refine. Similarly, the changes among textile prices are not consistent. Damask fell by 30 percent whereas the more delicate gauze and complicated three-shuttle cloth rose by 39 percent and 67 percent respectively. This divergence likely reflects differences in the cost of production, the latter two requiring more labor to produce than damask. Third, more prices rose than fell by twelve to five. Average the changes over this eighty-three-year period, and prices increased by roughly 50 percent. The per annum rate of 0.49 percent is consistent with the low end of the mild inflation one expects in a commercial economy.

The third set of prices draws on two sources recorded around 1562: prices that Hai Rui mandated when he drew up spending guidelines as a local magistrate in Zhejiang, and prices in the inventory of the confiscated property in Jiangxi province of one of the emperor's longest-serving ministers, Yan Song, who in that year was brought down at the age of eighty-two through his son on the doubtful charge of conspiring with Japanese pirates. The inventory of Yan's property was not publicly circulated at the time of the confiscation, but a private copy was printed much later under the title *Tianshui bingshan lu* (Heaven turning a glacier to water).[6] From Hai Rui's data and the Yan Song property inventory, I have selected fourteen items that appear to warrant comparison with items in the 1368 valuations list (see table 5.2 in appendix C, "Tables for Reference"). Three of Hai's items—armchair, iron wok, and knife—carry the same price in 1562 as they did in 1368. This lack of change in nominal price could indicate that prices for common manufactured goods did not rise on the gentle tide of mild inflation, which is to say that their real price measured against other things declined between 1368 and 1562. A second observation to be made is that while some prices fell, more rose than fell. Combining all the items in the table, the average price change over the two centuries from 1368 to 1562 is an increase of 31 percent, yielding a per annum inflation rate of 0.14 percent. Even if we narrow the sample to only those items for which the prices rose, the average rate of increase over two centuries is 79 percent, which when calculated as a per annum rate of inflation is only 0.30 percent, which is negligible.

Combining the results of these tables suggests that prices rose by roughly half a percent a year over the first century of the dynasty, when the new regime was actively rebuilding the economy. Calculated over the first two centuries, that rate declines to three-tenths of a percent. These findings seem to indicate that real prices rose during the first century of Ming rule but declined in the second century as the private commercial economy grew up alongside the state economy, lowering costs, speeding the circulation of commodities, expanding retail sales, and keeping wages stagnant.[7] This hypothesis may apply, however, only to manufactured goods and processed materials, as there is no evidence that real food prices declined in the second century of the Ming.

A third comparison with 1368 prices becomes possible by turning to the fourth data set, the prices that Shen Bang recorded in Beijing in the mid-Wanli era, roughly 1590. Given the density of Shen's records, a much more capacious comparison of prices is possible; in this case, seventy-two items. To control for the modest rise in nominal prices over those two centuries, I have converted nominal prices to real prices by indexing them to the price of a peck of rice. In 1368 the price of rice was 3.125 cents. No certain price exists for 1590. Shen paid several different prices, but for the purpose of the comparison I have settled on a price of 5 cents. The results appear in table 5.3 (see appendix C, "Tables for Reference"). The middle band of the table is reserved for the two principal grains that Shen's office purchased, rice and wheat. The prices in the two data sets indicate that the nominal price of both grains since

1368 rose by 60 percent. Above that middle band are listed thirty-one items of which the nominal price increased at a rate less than did the price of grain, which is to say that in relative terms their real price declined. Below the band are the thirty-nine items of which the nominal price increased at a rate greater than did the price of grain. Of this latter group, the real price of fifteen items increased by between 25 percent and 88 percent, nine items doubled their real price, and fifteen items more than doubled.

From these findings we may observe that the prices of everyday foods, excluding luxuries such as fruit and sugar, did not rise as much as did grain prices. In relative terms, they became cheaper. The real prices of tea, wine, and vinegar, however, doubled. Among textiles, other than notably expensive fabrics, the overall trend was that they became cheaper, as we might expect, given the intensification of commercial textile production through the sixteenth century. So too, the real prices of everyday manufactured goods came down. The greatest decrease was in the price of pepper. Pepper, however, is an oddity of Ming consumption. Not part of the Ming diet before 1368, it gradually pushed its way into Chinese cuisine as an exotic foreign commodity that Southeast Asian tribute envoys were required to present to the emperor, which he eventually off-loaded onto his officials in lieu of salary. What in the early Ming had been a much sought-after foreign import became a glut on the market later in the fifteenth century and was finally pushed aside by import substitution in the sixteenth. Pepper went from costly luxury to cheap everyday spice.

If we consider a few of the other commodities whose real prices rose more than did the prices of rice and wheat, wood and charcoal bear particular notice. The active deforestation of large areas of the Ming realm, especially in the early years of the Wanli era, was causing energy prices to rise: this may have been a factor affecting fuel prices. So too, the prices of almost all types of paper are higher. The demand for paper undoubtedly rose over the course of the dynasty as more official documents were required to be filed, more books printed, and more letters written. It appears that the demand for better-quality paper increased more rapidly than supply. In his record of local administration in the

1560s Hai Rui testifies to the recent escalation in the government's use of paper, to which he loudly objects.

> If my superiors were genuinely concerned about reducing expenses, they could reduce the numerous registers and documents and tally books required. In our communications, we could just use sheets that are plain and simple; no need for heavy and beautiful writing paper. Rather than binding statements submitted as evidence into a book, as was recently required, the documents could simply be put together in an envelope. Also, we could do away with the requirement that documents from separate bureaus be put in separate envelopes. And since the county is not far from the prefectural seat, we could make do with a single sheet of paper as an envelope and not paste them onto cardboard. Using standard formats for our documents and inserting them in registers would not only save paper but make it easier to read and consult them. There is no reason why we can't do all this. Every thousandth, even ten-thousandth, of a tael spent is extracted from the people's marrow. Save a thousandth, even a ten-thousandth, and there will be benefits all round.[8]

Hai had a knack of taking a sensible idea and inflating it into a panacea for ills that may not have been as serious as he thought, which is exactly what he is doing in this rant. Still, his candid opinion on the administration's excessive use of paper may be part of what explains why paper prices rose, especially in Beijing, where the government consumed more paper than in any other location in the Ming administration.

Returning to the middle band, the position there of rice and wheat emboldens me to suggest that the increase in their nominal prices—60 percent over 222 years—gives us a basis for calculating a compound rate of inflation through the first two centuries of the Ming of 0.21 percent per annum. This rate, low to negligible, is to be found in all preindustrial agrarian economies. This set of comparisons among the four benchmark sets of commodity prices serves, in a rough sense, to confirm Chen Qide's optimistic memory that prices early in the Wanli era were such that even poor people threw away the dregs left over from fermenting liquor, just as they gave the beans and wheat to their domestic animals rather than include them in their own diet. Severe inflation was not pushing food prices out of reach.

It was certainly not the case that "every household had all it needed" in the Wanli era, but given the insignificant rate of inflation that the foregoing comparisons suggest, it was not unreasonable for Chen Qide to remember the Wanli era as a time when most people could afford their daily needs. A crisis might arise, as it did in the late 1580s—drenching rains, then parched earth, then rice running to the intolerable price of 16 cents a peck—but the effect was of short term, and once it passed, people resumed paying the prices they had paid before it struck. Only those, like Chen, who outlived the Wanli era would be forced to discover that such good fortune could not go on forever.

Short-Term Price Change during the Wanli Era

The foregoing comparisons stretch across centuries. One price document survives, however, that allows us to focus on price changes over the short term, in this case, over one decade in the middle of the Wanli era. The annual ledger of Cheng's Dyeworks is remarkable in many ways, not least of which is that it is the only book of accounts of a Ming business that I have been able to track down, thanks to its having been reproduced in a set of historical documents preserved in Huizhou prefecture, famous for its wealthy native merchants.[9] The co-owners of the dyeing workshop were natives of Huizhou, which is why the ledger ended up there, but it appears that the workshop was located in Songjiang prefecture in the heart of the cloth-producing region around Shanghai.[10] Composed of forty-two folios folded and bound into a book of eighty-four pages, the ledger consists of annual year-end summaries of the firm's assets and profits.

The first document in the ledger is dated the first day of the fifth month of the nineteenth year of the Wanli era (21 June 1591). It lists the initial shares paid into the firm by two men, Cheng Benxiu and Wu Yuanji, in the amount of 2,417.412 taels. The next document, dated the first day of the seventh month two years later (28 July 1593), is the first of the annual statements that follow. The list for 1593 is the least detailed of the annual reports, and the statements for the years 1595 and 1596 are missing. In 1597, the year end for accounting purposes was moved from the first day of the fifth month to the sixteenth day of the third month, though later it reverts back to the fifth month. The format varies slightly from year to year, and the handwriting changes between 1601 and 1602. Otherwise, the presentation and information are reasonably consistent from one year to the next. The final document in the ledger is an agreement appended to the 1604 account signed by Cheng Guanru and Wu Xiajiang, Wu Yuanji's younger brother, on how to dispose of Wu Yuanji's share in the business following his death the previous autumn.

The ledger was kept in order to record the company's assets on an annual basis. Each year's report lists the goods in stock, their quantities in one row and their total values in another row below this. The order of entries changes somewhat from year to year, with processed materials generally preceding raw materials. These are valuations rather than market prices, though we can assume that the values had to be close to market prices if the co-owners were to agree on their value. What is striking, and what may support this assumption, is that the per-unit value of almost everything changes every year. Table 5.4 (see appendix C, "Tables for Reference") displays the values of commodities for which the ledger reports at least three valuations between 1594 and 1604: indigo, cloth, distillery dregs (purchased as a source of potassium, which was needed to bleach cloth and enhance dye quality), lime (calcium hydroxide, used for fixing dyes), firewood, rice, and woven rush mats (used to wrap bolts of dyed cloth), plus two items that I can translate but not construe, "cloth heads" and "shaved skins."

Consider the price movements of each category, starting with indigo. Of the eight types of indigo the ledger lists, three appear in at least three of the annual summaries and so are listed in table 5.4. Local indigo appears twice at 1 cent per catty in 1594 and 1597, then at 0.72 cents in 1598. Rizhang (probably a firm name) indigo is priced in 1597 at 1.53 cents. Courtyard indigo registers its highest price of 1.3 cents in 1600, then falls to 1.1 cents in 1601 and 1603. The only observation that can be drawn from these few data is that indigo prices slipped slightly between 1600 and 1601.

The data for cloth are more abundant. The price of blue (that is, dyed) cloth is highest in 1597 and 1604, though even that observation is troubled by the fact that there are two separate entries for dyed cloth in 1597, one priced at an all-time high of 22.9 cents per bolt and the other at an all-time low of 19.6 cents, with no indication of why two types were reported for that year. The price of white (that is, undyed) cloth hits a low of 14.4 cents per bolt in that same year of 1597 and goes to a high of 18 cents in 1602, quite out of line with the prices for the other years. Undyed cloth from Gangshang (again, I presume, the name of a firm with which the dye shop did business) reaches a low of 14.7 cents in 1599 and a high of 16 cents in 1604. Bean-fiber cloth shows less fluctuation, its low of 14.9 cents in 1603 followed by a high of 16.4 cents the following year. Cloth dyed at other workshops is much cheaper than the other types of cloth, reaching a high of 6.5 cents in 1600 and a low of 4.9 cents in 1603. In this category I have also included the two valuations in the ledger for cloth dyed for book covers in order to show that it was possible for some prices not to move. That cloth held steady at 20 cents through 1600 and 1601, the only two years for which it was reported. If we step back from these data and look for a general observation on the price movements of cloth, none suggests itself. Each item's price moved independently of the prices of the other items.

Shifting to raw materials, distillery dregs reached a high of 12.5 cents per jar in 1594 and a low of 9.5 cents in 1601. Firewood moves from its lowest price of 1.1 cents per catty in 1600 to its high of 1.5 cents the following year, then two years later goes down again to 1.1 cents. For lime there are only two prices, 3 taels per *fang* (literally "room," the unit in which it was measured) in 1598, 1601, and 1603, and 3.3 taels in 1599, 1600, and 1602. In light of what I have found in other sources, the rice prices are unusually high, rising from 0.6 to 0.7 taels per hectoliter between 1593 and 1599, falling to 0.55 taels in 1601, then rising to 0.75 taels through 1602–3. Sheets of wrapping material woven from reeds slips from 0.25 cents in 1600 to 0.23 cents in 1604. "Cloth heads" sink slightly to 9.7 cents per catty in 1600 then rise to 11.9 cents in 1604. "Shaved skins" reach a low of 0.143 cents each in 1601 and a high of 0.17 cents in 1603. Cumulatively, the movements of these prices cannot be unified into an overall price trend, nor is it apparent that there is any consistency across similar types of goods. Prices were slightly higher in 1604 that in any of the previous ten years, though the difference is so marginal as to be insignificant.

What do these short-term price movements in the mid-Wanli era tell us? The first thing is that prices moved from year to year, usually in very small increments. The second is that prices did not move in unison, which indicates that price fluctuations had to do with conditions that were particular to the goods as priced within that economy rather than that prices as a whole were being moved by larger economic forces. What these two observations indicate is that no overall price movement can be detected through the mid-Wanli era. By way of demonstration, take the first and last prices in each category of table 5.4 (again, see appendix C, "Tables for Reference"): the price of indigo falls, as do the prices of distillery dregs, firewood, and rush wrappings; the price of lime remains the same, as does the price of book cloth; and the prices of all other types of cloth rise by roughly 6 percent. The one instance of significant price change is rice, the price of which rises 25 percent, though as already noted, I am not confident how to interpret these rice prices, especially as the years 1603 and 1604 were warmer and well watered.

Viewing the dyeworks prices as a whole, I have to conclude that there is no evidence of any short-term change, regardless of what factor we might be disposed to identify as the source of such change. This finding robs us of any argument for identifying evidence that would support the second stage of Fischer's wave model, in which prices, after a long, slow rise, ramped up preparatory to the sudden surge to come in the third stage. No wave can be detected in the mid-Wanli era, and no preparation for the huge surge that would follow in the Chongzhen era. Accordingly, we have to conclude that the Chongzhen Slough was not the destination toward which the Ming economy had been heading, either since the start of the dynasty or in the decades immediately preceding. I therefore find no reason to build backward from the Chongzhen price chaos to argue that inflationary pressures internal to the economy were at work. The Chongzhen surge rose on its own. Something else was afoot.

Price Restabilization after the Chongzhen Era

The price history of the Qing, a subject in its own right, is beyond the research range of this writer. But I shall dip into the evidence of price movements in the new dynasty's first half century through yet another participant-observer of the Chongzhen Slough on the Yangzi delta.

Ye Mengzhu was born into a lesser gentry family southwest of Shanghai. The family maintained an urban residence in the prefectural city of Songjiang, but their rural home lay in the countryside between Songjiang and Shanghai. In his memoir *Yueshi bian* (A survey of the age), Ye looks back from the 1690s to reconstruct the history of the region from his childhood in the Chongzhen era down to 1693, the latest date that appears between its covers.[11] Carefully organized and precisely written, the book reads as something like an ethnography of the Yangzi delta through the first five decades of Manchu rule.

Like his contemporaries, Ye was fascinated by the movement of prices, to which he devotes an entire chapter, the seventh. The title, "Shihuo" (Food and goods), imitates the title of the section of the dynastic histories that reported on fiscal matters such as population, production, and taxation. Ye's interest is not in population or taxes, but purely in what things cost. He opens with a general statement on the subject: "Commodity prices are not stable, and it has been this way since ancient times. Who would have imagined that over the past thirty-odd years, the price of a thing could reach tens, even hundreds, of times what it had been? Not only that, but what was expensive could become cheap, and what was cheap could become expensive, changing unfathomably."[12] Just as Chen Qide did, Ye starts his account of the prices he experienced in his childhood by placing himself at his grandfather's knee, using this as a point of reference from which to comprehend the wave of price confusion through which he lived. In his case, as we are about to see, the arcadian simplicity that Chen Qide thought he could recall back in the Wanli era was not there in the Chongzhen era for Ye's having.

Ye Mengzhu starts his tale of woe in 1630, "a famine year when grain was expensive" and soup kitchens were opened to feed the famished. "I was still young at the time," he writes, "and knew nothing about commodity prices." He dates his first prices to 1632, "when white rice cost 120 coppers per peck, or in silver, 10 cents. The people suffered at this price, so you can imagine what the price of rice in 1630 must have been like. Then autumn came, and the price of early rice was only 65 or 66 coppers. Thereafter the price of rice stayed around 100 coppers." As we know from the previous chapter, the standard famine price through much of the sixteenth century had been exactly this price, 100 coppers per peck. As of 1632, the famine price of rice had become the everyday normal price. Ye attributes some of the erratic movement of rice prices to the bewildering fluctuation in the exchange rate between copper and silver as counterfeiters adulterated copper coins by melting them down and recasting them into cheaper versions of themselves.[13] Some scholars have argued that the exchange rate was disturbed not by the falling value of copper but by the rising cost of silver due to the decline in imports from the Americas in the 1640s, though Ye has nothing to offer to that debate.[14] A more historically grounded way to think about counterfeiting might be to treat it as a way of making more coins available in a local economy in which they were in short supply, especially at a time when those who had silver withdrew it from circulation in order to save it against an uncertain future.

Like Chen Qide, though in a slightly less moralistic way, Ye Mengzhu looked for lessons from the difficult times he had lived through. "No generation is without the warnings that disasters and auspicious signs give" is how he begins his opening chapter. "The disasters recorded in the registers of history are too many to be counted. As for the worst I have seen, nothing could match the drought of 1641." This was Ye's annus horribilis, as it was Chen Qide's, beginning with a summer drought followed by a plague of locusts and ending in a massive famine. Officials approached local magnates to set up gruel kitchens, though many of the famished dropped dead along the roads on the way to reach them. Ye then turns to prices. "This was the time when hulled rice was 5 taels per hectoliter, and beans and wheat were only a little less. Even chaff and bones were suddenly expensive. A guest who dropped by and was fed a simple meal regarded it as a sumptuous feast. Laborers could be hired for nothing but the food to fill their bellies, even if it was nothing but barley." Half of Songjiang's population depended on textile weaving for their livelihoods, so when the buyers did not show up, "they sold their children for a meal, and butchered corpses to grill." A military requisition of wheat that the court added to Songjiang's tax bill to sustain the troops holding the northern border against the Manchus drove the local price of grain above 5 taels. A member of the local gentry persuaded the government to lower the commutation on the wheat tax to a more favorable rate of 1.5 taels per hectoliter. The summer wheat crop survived long enough to be harvested, which allowed some people to meet the tax burden. "Then an epidemic broke out, going from one house to the next, bringing death to each in turn. Of everything that I have seen since my birth, this was the absolute worst year."[15]

Ye then moves forward to the summer of 1644. A six-month-long drought inspired two new local sayings. One was, "The price of rice is expensive, but the price of water is twice as expensive." The other was, "The famished wish to die, but the parched wish to die even more." The 1644 drought caused the entire economy of Songjiang to collapse. "No merchants took to the roads, and commodity prices soared," Ye wrote. The crisis finally began to ease at the end of that year, but recovery was gradual. "Only when we got to the twelfth month [January 1645] did rain fall for several days in a row. It was enough to quickly moisten [the soil], yet the people were utterly exhausted."[16]

Ye continues his story of price fluctuation beyond the fall of the Ming down to the 1690s. It was a bewildering story to tell, for price movements were not unified, obliging him to narrate these movements product by product. To simplify his report, I limit the following account to eight of the three dozen commodities he followed, starting with rice.

Rice climbed to half a tael per peck in 1642, subsided in the 1650s to a range of a quarter to two-fifths of a tael, then slid down below a fifth of a tael in the 1670s, ending up at 8 to 9 cents in the 1680s. Wheat climbed to only a quarter of a tael per peck in 1642. By the 1670s its price was in the range of 12–13 cents. Wheat then moved down in the 1680s to match the price of rice, at 8–9 cents. Soy beans rose from 5 cents in 1640 to 35 cents in 1661, but by 1682 were down to 6–7 cents. Pork, which like grain is something of a bellwether commodity in the Chinese market, had been 2 cents in Songjiang in the 1620s, which is the price for which pork had sold since the beginning of the dynasty. It had been driven up to a high of 12 cents in 1645, restabilizing at 5 cents in 1680. Sugar, which had been priced at 3–4 cents in the 1620s, leapt tenfold to 40 cents in the 1640s. It sank to 2–3 cents in 1681, finally recovering and settling at 5–6 cents in 1690. Firewood had been 6–10 cents a bundle in 1620. In 1646 it was 50–60 cents, though eventually it restabilized at 12–14 cents in 1688. The only commodities that returned to their 1620 price were textiles, though raw cotton took an unusual route to get there. Priced at 1.6 taels per 100 catties in 1620, it soared to 4–5 taels in 1628, then subsided and rose again in 1649, though this time only to 3.5 taels. By 1684 it was down to as low as 1.3 taels. Cotton cloth had been 15–20 cents per bolt in 1628, rose to 50 cents in 1654, and by 1690 was back to 20 cents.[17]

Ye Mengzhu's prices do not march in step with each other, though his reports do lead to the overall impression that prices across the board in Songjiang were already higher than normal in 1620, and that they rose sharply over the next twenty-five years, reaching levels two to ten times higher than they had been in the 1620s. In every case, after the mid-1640s, they subsided. A few, such as for cotton cloth, returned to their 1620 prices. The prices of most other commodities did not restabilize to that level but settled at a new level that was roughly twice what they had been before the Chongzhen era. Ye's evidence thus indicates that the price regime in the 1690s did not return to the price regime that had prevailed in the 1620s. Nor can these prices be used to describe anything like an equilibrium. The prices of most things had doubled and were not immune to further movement.

The Chongzhen Price Surge

How might Ye's testimony help us to analyze what happened to prices during the Chongzhen Slough? To consider this question, let us return briefly to the full Ming record of famine grain prices. If I cover ground already covered, I do so to help situate what we might choose to call the Chongzhen price surge.

Just as prices in general rose gently from the fourteenth to the seventeenth century, so too the famine price of grain moved upward. A famine price of 100 coppers was fairly standard through the first two centuries of Ming rule. In brief moments of intense crisis, the famine price went above that ceiling, but in every case the price of grain returned to a level close to what it had been, leaving the overall price regime unaltered. Figure 5.1 shows how the famine price of grain in copper moved up from that floor of 100 coppers to 1,000 coppers during the 1630s. Silver prices rose as well, though not in quite so clean a pattern. In 1428, 10 cents was considered a famine price. By the 1480s, famine prices had moved to a level of about 15 cents, climbing still further through the sixteenth century. In the context of the ordinary price floor from which they strayed, famine prices found a ceiling that loosely echoed the modest rate of inflation in the normal price of grain, on the order of 0.2 percent per annum until the mid-1500s.

Once into the Wanli era, the scale and intensity of the First and Second Wanli Sloughs, which exceeded earlier sloughs, produced far more disturbance among famine grain prices. The old copper price of 100 coins continued to be reported as the famine price of grain in some areas as late as the 1610s, though that limit was broken and largely discarded during the Second Wanli Slough. Still, these prices did not disrupt the Ming price regime to any significant degree. Despite growing pressure on grain prices, they largely returned to normal after the crisis had passed. Only when we get to the wild destabilization of the 1630s and 1640s—as the conditions under which farmers found themselves struggling to grow enough grain to feed everyone degenerated—do disaster prices move into an entirely new register. The logic underpinning famine relief policies up to the Wanli era had been that prices would eventually return to normal, and that the role of the government was to nudge the arc of change back to prices as they had been before the famine. Famine relief was supposed to tide the people over until agriculture could revive and restore food production and reinstate prices as they had been. During the later years of the Chongzhen era, however, earlier

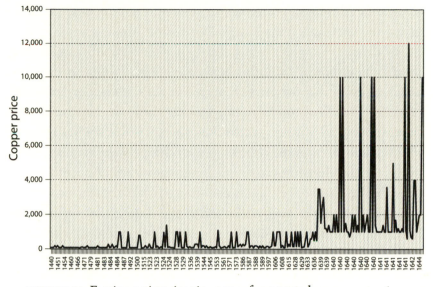

FIGURE 5.1. Famine grain prices in copper for reported years, 1440–1647.

price levels shattered. Each exceptional price that towered into view was only dwarfed by the next to arrive, and confidence that the previous price had been an extreme beyond which there could lie nothing worse faded away. Famine prices soared into the stratosphere to levels that attested that the famine price of grain appeared to have no ceiling whatsoever, and no remedy could be found.

The story of Ming prices does not end with the collapse of the dynasty, as our review of Ye Mengzhu's price records in his *Survey of the Age* has shown. The price chaos at the end of the Ming continued on past 1644 as China slid into the Maunder Minimum, which continued to pressure agriculture and force prices through the early decades of the Qing. When the price of grain partially stabilized in his early adulthood, it was in the range of 2 to 3 taels per hectoliter, a price that would have been unimaginable at any earlier time. It took decades for the price finally to drop, but the level at which it came to rest was the level of early Chongzhen prices, which were already well above what had been the norm for grain prices in the Wanli price regime.[18]

Where we locate the floor from which prices moved affects how we interpret the shape and amplitude of that movement. Some historians have taken Ye Mengzhu's account of the decline of price changes as evidence of what is called the Kangxi Depression. There is much qualitative evidence to support the fact that these were difficult years, as other scholars have shown.[19] Still, this analysis arises from a Qing perspective, which assumes that the prices reported in the Chongzhen era can be taken as a baseline for price movements through the seventeenth century. From a Ming perspective, Chongzhen prices are not a reasonable baseline for reconstructing long-term change. They give expression not to the acceleration phase of a long-term wave (Fischer's fourth stage) but to short-term climate conditions that played havoc with the Chinese economy through to the end of the Chongzhen reign; that, at least, is my interpretation. To take Ye's early Chongzhen grain prices (10 cents per peck in 1632) as normal, as Qing price historians have done, is to be blind to the fact that the price of grain under other than famine conditions mostly stayed below 5 cents. Starting the clock in the Chongzhen era rather than a few decades earlier obscures the peculiarity of prices during that era and misses what actually happened.[20]

To adopt a Ming rather than a Qing perspective, what Ye Mengzhu's prices show is that prices restabilized toward the end of the seventeenth century, not to half their earlier level (which is what they appear to do if we take the prices of the 1640s as the benchmark), but to twice that level. Not all commodity prices moved in the same way, especially raw materials such as cotton, which relied on transportation and longer-distance commercial networks for their financial viability and were dampened by labor costs.[21] If Ye's price data suggest that prices through the seventeenth century moved in a long wave, that wave subsided to a price level well above where it started. Rather than a wave, what Ye's memoir shows is that the famine prices of the Chongzhen era caused a surge that did not subside nor return to a familiar equilibrium. The prices at the end of the seventeenth century were not normal, but a new normal. These new normal prices altered costs and eroded standards of living, placing such severe strains on household economies that those who lived through the Kangxi Depression could barely earn what the new prices required. In that sense, people of the early Kangxi did go through a depression, but one created by the price effects of the

Chongzhen Slough. Beyond that, the new normal was not so much an equilibrium as it was a temporary pause, for prices would continue to rise, roughly in tandem with population, through the course of the eighteenth century.

Famine Prices and the Role of Climate
in the Fall of the Ming

If famine prices from the Chongzhen Slough and beyond describe a shape, it is more a tsunami-like surge than a wave that rises and subsides. To push prices as high as it did, and in such short order, a surge of this sort had to have been caused by a powerful external force: not the slow-acting effect of inbuilt inflation or changes in the money supply, but the sudden shock of profound climate disruption.

We now know that the crisis of the late Ming was part of a global crisis.[22] There was a difference at the two ends of Eurasia, however, and comparing them not just as climate zones but as human societies helps to clarify what was particular about the burden of the Little Ice Age on Chinese agriculture. Emmanuel Le Roy Ladurie concludes his massive survey of European climate history by observing that, "taken as a whole, it is really excessive precipitation which is by far the main danger for temperate climates such as ours in the middle of western Europe north of the Mediterranean." He regards wet and dry as the key register of climate variability, with wet as the more threatening pole, even going so far as to suggest that "dryness is rather a good thing, at least so long as it is not accompanied randomly by heat or when it becomes excessive. As for cold, notably in winter, its effects are mitigated. It all depends on its modus operandi (with or without snow, for example) as to whether it is for good or for ill."[23]

Ming China presents a somewhat different case. Participant-observers such as Ye Mengzhu and Chen Qide blamed drought as the greatest threat to food production. They and everyone else believed that drought was what elevated price to intolerable levels. They were certain that the only factor that could bring down the price of grain was rain.

Physical proxies of climate change in this period suggest, however, that cooling—the climate signature of the Little Ice Age, after all—rather than drought was the more severe factor driving the Chongzhen crisis.[24] This is not a distinction that is easily detected in the price records on which I have relied for this study. Gazetteer editors sometimes include a brief note to explain why prices soared, yet almost never do they attribute a famine grain price to falling temperatures, except when they could see the physical evidence of ice-choked canals or snow or frost descending on their fields out of season. This omission was due in part to the lack of a standard metric for warmth, as well in part to the difficulty of detecting the effect of cold on plant growth until well past its onset. The brutal evidence of drought provided all the explanation anyone seemed to need.

The difference between cold-and-wet conditions in Europe and cold-and-dry conditions in China deserves close attention as well. It matters because European and Chinese systems of agriculture evolved in ways that differ in their capacities to respond to precipitation stress. As Le Roy Ladurie has observed, European food production was better positioned to respond to cold so long as it was not accompanied by excessive rainfall, at which point harvests failed. This resilience derived in part from the larger place of animal husbandry in the European diet as well as from the use of more drought-tolerant grains. When cold struck China, by contrast, the more dangerous combination than cold and heavy precipitation was cold and drought. Both wheat and rice are notable for the volume of water they require to grow, yet rice needs significantly more, requiring roughly 2,500 liters to produce one kilogram of grain as opposed to roughly 1,500 liters for wheat. When rice is irrigated, as it usually is, the water that it can be expected to absorb amounts to about two-fifths of the water in the system. As a result, rice agriculture not only uses more water but contributes to water scarcity. Rice is a good crop to cultivate when the challenge is too much precipitation, as paddy fields are constructed to drain and can be more readily made to release standing water than can dryland fields, but its demand on water is less tolerant when drought strikes, especially suddenly and for months on end. Worse is when colder temperatures shorten the growing season, intensifying the impact of water shortage. During phases of extreme climate distortion, when a spring or summer might pass without a drop of rain falling for months on end, drying the fields and emptying the canals, nature, in our language, and Heaven, in theirs, pushed farmers beyond endurance.

As the Ming slipped further into Chinese historical memory, the shock of the Chongzhen crisis tended to be forgotten, and the burden of catastrophe shifted to human rather than natural forces. By the eighteenth century, many a commentator was moved to look back into the Ming and cast it as a better time. The editor of the 1733 prefectural gazetteer of Yangzhou, which lay across the Yangzi River north of the delta, looked past the seventeenth century to the sixteenth to fancy it as an innocent time when, unlike his own troubled era of steeply rising prices, "prices were modest and customs plain."[25] If the Ming was this lost arcadia, grain prices had to tell the story as a cautionary tale against the crass commercialization and the abandonment of the old ways of his own time. For this eighteenth-century author, as for every one of our observers, the explanation had to be a moral one. Morality was the only ground on which it was possible to imagine the economy systemically, which is to say, in terms higher than those the price-shocked confronted. Climate, like Heaven, was essentially a system beyond analysis.

From our perspective, the move some three decades ago to direct attention to the global trade in silver in the sixteenth and seventeenth centuries served to enlarge the historical recognition of China's place in the world, freeing us from the political narrative of moral decline. To exploit the findings of global history in order to reexplain the crises and the collapse of the Ming dynasty without factoring in climate, though, misses what has been hidden in plain sight, and that is China's vulnerability to the Little Ice Age in its most extreme phase. The silver that washed into China through the Wanli to Chongzhen eras may have had an impact on certain sectors. But agricultural production remained the basis on which the Ming economy rested, the means by which it fed an expanding population, and so a shock to that production could be deadly. When grain in the fields everywhere wilted from drought or perished from cold, the impact was not just human hunger but hunger prices, which is why famine grain prices offer one of the best documentary proxies we have for an environmental history of the dynasty. The strain that cold and drought imposed on grain production during the very worst years of China's Little Ice Age would not be matched again until the 1850s.[26] The Ming price and political regimes could not withstand the utter collapse in food supply. It may be that the Manchus were better adapted to colder, drier climate as collapsing temperatures pushed them southward into Chinese territory to occupy that realm at its most disordered and to reshape it in ways that continue to resonate today. The people of the Qing managed to adjust to post-Chongzhen prices and move eventually into a new price regime in the eighteenth century, until the climate disruptions at the close of the Little Ice Age in the mid-nineteenth century drove them on to the next wave of famine, civil war, and, in its turn, dynastic collapse.

Climate and History

WHEN I BEGAN collecting references to prices in Ming documents, I had no idea where the data would lead me. I fondly hoped that it would be toward some sort of empirical certainty about what it cost ordinary people to live during the Ming period. While I have unearthed many prices that real people paid to buy real things, that hope proved elusive. The closest I have come to contributing to historical knowledge of this sort, albeit with only referential proof, has been to propose, first, that a Ming household of the poorest sort needed a little over 14 taels to get through a year, while the cost of living for a respectable household, to use the language of European history, was just over 23 taels; and second, that the annual wages of the poor were between 5 and 12 taels, while respectable wages were up between 14 and 22 taels.

These generalizations far from summarize the findings of this project. More surprising, and more intellectually engaging, has been where the search for price data has led me, to environmental history. The fall of the Ming dynasty has traditionally been narrated as a period of political factionalism, failed administration, dwindling tax revenues, and rural rebellion, all of which has been shrouded by the larger judgment of moral failure. This narrative—that the Qing Great State was able to invade and occupy Ming territory only because the Ming failed—is exactly that: a story put in place to legitimate the dynastic transition from Ming to Qing. Those who authored, and authorized, this narrative were none other than the conquerors. Dorgon, the commander of the invading Manchu army, summarized events by allowing that "Emperor

Chongzhen was all right" and insisting that those who should take the blame were his military officers, who "were of bogus merit and trumped up their victories," and his civil officials, who "were greedy and broke the law." These morally flawed characters allowed Dorgon to conclude confidently, "That is why he lost the empire." The same judgment was inscribed on the mortuary tablet at Chongzhen's tomb outside Beijing. There it states that the emperor "sacrificed his life on behalf of the nation" whereas those around him "lost their virtue and let their country perish."[1] Moral failure can be a comforting tale for those who get to push aside the failed, as well as for those who regard themselves as powerless to do anything about their failure. Many a Ming loyalist in the early decades of the Qing embraced the judgment, indulging in cries of "mea culpa" and clutching at whatever bits of moral wreckage allowed them to account for why they had outlived the dynasty to which they had proclaimed their loyalty—and in doing so, won the indulgence of their new masters.

It is not helpful to deny the impact of the failures of judgment throughout the Tianqi and Chongzhen administrations. Had these administrations been in the hands of competent officials devoted to more than advancing their own factional and personal interests, it is plausible that some of the fiscal and military crises could have been avoided. In light of what local sources reveal of the people's suffering toward the end of the dynasty, however, the weight of explanation for the misery might better be made to go in the other direction, not by measuring the actions of individuals against absolute moral standards, but rather by setting their actions in relation to the desperate conditions of the Chongzhen era. Officials during earlier sloughs found themselves up against similar difficulties when harvests failed, food disappeared from markets and state granaries, and prices rose. The scale of the final slough, however, was beyond what anyone at those earlier times could have imagined. Some sloughs ended in political disaster, such as the palace coup against Emperor Jingtai in the winter of 1456, but most dissolved back into the fabric of what had been there before. What marked the Chongzhen Slough was the shocking level to which grain prices rose at a time when a well-organized military force across the northern border, burdened by its own food shortages, was waiting for the moment to strike. To blame the collapse of the dynasty on the moral failings of Chongzhen's officials seems almost beside the point, given such conditions. True, many an official from top to bottom took refuge in whatever opportunities came his way for personal enrichment and safety at the expense of the public good, but most did so in the face of conditions that even contemporaries could not have imagined.

Setting the events in this book in the context of the Little Ice Age brings to the fore the climate context that the people of the Ming and their regime faced through the fifteenth to seventeenth centuries. Grain prices were the device by which the relationship between solar energy and human demand was mediated. The surges in these prices during the five environmental sloughs from Jingtai through Chongzhen, each of which pushed prices ever higher, persuaded me to adopt this broader frame. It was simply impossible to ignore that what determined the experience of Chinese during these centuries at the most fundamental level was their relationship with nature. When an economy depends on solar radiation as its source of energy, nature—writ as large as you like—must be recognized as the factor determining the viability of society or state. It was nature that set the limits of the possible, to invoke Fernand Braudel's phrase one last time. Those limits were not necessarily absolute. Frequently throughout history, people have intervened to shape environmental conditions in their favor. These responses have included, among other initiatives, building infrastructure (such as irrigation and drainage canals), mutating crops (such as early-ripening rice), developing institutions (such as granaries and grain markets), devising technologies (such as water pumps), and controlling rates of reproduction so as to limit demographic growth and ease the pressure on food supply.[2]

So what might the record of Ming prices tell us about these potentially anthropocenic strategies? To put the question more harshly, was the Ming conspicuously poor at responding to climate change? I would respond to this question by challenging its buried assumption that human resilience operates autonomously of the conditions eliciting it. It is necessary here to distinguish between long-term and short-term climate disruptions, between climate change and the weather, to put it another way, although at the risk of burying the close link between the two. Climate historians took up the concept of the Little Ice Age to capture a long-term trend in human history, initially on the basis of the European record, though now as well on the strength of what Asian climate proxies reveal. Historical research shows that people, including the people of the Ming, regularly intervened through the Little Ice Age to ameliorate their environmental situations. In all five of the ways just noted (infrastructure, genetics, institutions, technologies, and demography), Chinese have been conspicuously resilient in their response to environmental stress. These innovations were not accomplished overnight, nor were they always able to compensate when the pendulum of disruption swung wider than before, but adaptations were developed.

Short-term disruption, however, has a different human impact. Whereas long-term disturbance forces humans to adapt their practices to adjust to new circumstances, short-term catastrophe, especially when it is sudden and intense, is more likely to overwhelm adaptation than to stimulate it. This, at least, is how I interpret the price spikes that I have detected from the Jingtai era to Chongzhen, when grain prices rose far above anything people had previously experienced. The sources from which I have scoured these famine prices have almost nothing to say regarding adaptation. Far more often they simply note that the impact was mass starvation and that there was nothing anyone could do.

To the extent that there was resilience, it was happening along a different, longer time line. We have seen some evidence of this adaptive capacity in relation to famines that did not descend into mass starvation. Take the case of the Henan famine of 1594, in which Emperor Wanli and his consort directly intervened. When the hectoliter price of grain rose to 5 taels, the official in charge of relief exploited this price to incentivize private grain merchants to load up their grain barges and send them down the Yellow River into the affected region. There might have been a time when government granaries could have supplied that grain. After all, the dynastic founder mandated the construction of granaries throughout the country. The costs of keeping granaries in good repair and stocked with grain were not negligible, however, which served to undercut their maintenance. Unwilling to do what was necessary to stem the decline, the central government gradually gave in to their deterioration through the fifteenth century and in 1527 reduced the quotas on the amount of grain that should be stocked to a level insufficient to meet a severe subsistence crisis.[3] Officials on the ground understood what they were up against. The 1594 famine was averted not because the market intervened in some abstract sense, but because the official in charge knew that he did not have access to state granary stocks sufficient to meet the emergency, and that the market was where to go to meet the need for grain.

The difference between the Wanli and Chongzhen Sloughs was not that knowledge of how to combine state and private commerce had been lost, or that officials had retreated from innovation. The core difference was scale. The cold, dry climate of the late 1630s had a sustained impact on grain production far exceeding what the climate had imposed previously. By that point, there were no sources of grain elsewhere that could be tapped, either by the government or by the market, to meet shortfalls in food supply. It is undeniable that the military disorder of the late Chongzhen era caused breakdowns in security and communications that exacerbated the impact of food shortages, but that impact would have had no purchase had there been adequate harvests, which there weren't. In highlighting the events of the Chongzhen era through the last two chapters, I have stepped back from the long-term effects of the Little Ice Age through the last two centuries of the Ming dynasty to focus on the short-term impact of the global crisis of the late 1630s and 1640s, when climate disruption undermined the viability of the Ming regime. Rather than count up the human failings that intensified this impact, we come closer to the reality of the collapse by looking first at climate.

Shifting the scale of analysis from long to short term does not negate or override the role of the long-term cold phase in which the people of the Ming found themselves, and to which they adapted; nor does it slight the capacity of human agency to mitigate its effects. It is to propose that whatever resilience the people of the Ming were able to muster as the Little Ice Age lowered the temperature floor of their annual agrarian cycle, it was insufficient when that floor collapsed, as it did so conspicuously between 1638 and 1644. Whether China under a different regime might have withstood the environmental calamities of those years may be an instructive counterfactual, but that was not what happened. If the ghost of environmental determinism lingers just outside my analytical door, it is not a ghost I am prepared to deny. Many hands were on deck when the Chongzhen Slough struck, not all of them incompetent, but they were as overwhelmed by the scale of environmental stress as we have been inattentive to it until recently.

I will end this book with a Ming voice from the south end of Shaanxi province, the testimony of someone who had probably been sent there as a local official. His identity is now lost to us, but his words survive because he had them carved in stone. That stone stele used to stand at the outer corner of the wall surrounding a temple three kilometers south of the city of Huazhou, which lies in the Wei River valley to the east of the ancient capital of Xi'an just before that river flows into the Yellow River. The stele got knocked over during the armed struggle against the occupying Japanese army in the Second World War. After the war, rather than put the stone slab back, someone had the bright idea of using it to shore up the side of a collapsed well. Who needs a historical artifact when a well can be repaired and brought back into use? Archaeologists recovered the stone a decade later, removed it, and gave it a home in the provincial museum, where it stands today.

The text on the stele bears this title: "A Record of Sorrows in Response to These Times."[4] The author's name was placed at the end of the text, but that part of the stone has been chipped off, probably when it fell during the war. Fortunately, the single most important fact about this inscription is still legible, and that is the date, 1643, the year before the Ming collapsed. The text begins by describing the harrowing experience of living through those terrible times. The author captures the extremity of the experience, just as Chen Qide did, by turning to prices. After a four-line poem coaxing tears of sorrow from his readers, he expresses the horror that was 1643 with the most shocking data he could muster, a price list:

> 1 peck of rice or millet, 2 taels 30 cents
> 1 peck of wheat, 2 taels 10 cents
> 1 peck of barley, 1 tael 4 cents
> 1 peck of buckwheat, 90 cents
> 1 peck of beans, 1 tael 80 cents
> 1 peck of bran, 50 cents
> 1 peck of chaff, 10 cents.

In the past, the author explains, famines might drive the price of a peck of grain to 30 cents (which rarely happened before the mid-sixteenth century) or even 70 cents (which is not a price attested with any regularity until barely fifteen years before 1643). To people living in the Wei River valley in 1643, such prices were now already history. The past had vanished, the present was inconceivable, and the future lay beyond comprehension. They had no inkling that the Manchu invasion, the Maunder Minimum, and the rest of China's history was about to begin, at their cost.

Units of Measurement

TABLE 1.1. Units of Measurement

Unit		Translations	Metric equivalent	Imperial equivalent
Money				
liang	兩	tael	37.3 g silver	1.3 oz.
qian	錢	mace	3.73 g silver	
fen	分	silver cent	0.373 g silver	
wen	文	copper		
Volume				
shi	石	hectoliter	107.4 liters	23.6 gal.
dou	斗	peck, decaliter	10.74 liters	2.36 gal.
sheng	升	liter	1.07 liters	0.94 quart
Weight				
dan	擔	picul	59.68 kg	133.3 lb.
jin	斤	catty	596.8 g	1.33 lb.
liang	兩	tael	37.3 g	1.3 oz.
Length				
cun	寸	inch	3.2 cm	1.2 in.
chi	尺	foot	32 cm	1.26 ft.
zhang	丈		3.2 m	3.5 yd.
pi	疋 / 匹	bolt: 32 chi (Hongwu era)	10.24 m	11.20 yd.
		bolt: 37 chi (Jiajing era)	11.84 m	12.95 yd.
		bolt: 42 chi (Wanli era)	13.44 m	14.70 yd.
Area				
mu	畝		0.066 hectare	0.165 acre

Sources: Boxer, *Great Ship from Amacon*, 181; *Songjiang fuzhi* (1630), 15.3a–b; Schäfer and Kuhn, *Weaving an Economic Pattern*, 29, 38; Wu Chengluo, *Zhongguo duliangheng shi*, 54, 58, 60.

Reign Eras of the Ming Dynasty, 1368–1644

TABLE 1.2. Reign Eras of the Ming Dynasty, 1368–1644

Emperor's personal name	Reign title	Dates
1. Zhu Yuanzhang	Hongwu	1368–1398
2. Zhu Yunwen	Jianwen	1399–1402
3. Zhu Di	Yongle	1403–1424
4. Zhu Gaozhi	Hongxi	1425
5. Zhu Zhanji	Xuande	1426–1435
6. Zhu Qizhen	Zhengtong	1436–1449
7. Zhu Qiyu	Jingtai	1450–1456
8. Zhu Qizhen	Tianshun	1457–1464
9. Zhu Jianshen	Chenghua	1465–1487
10. Zhu Youtang	Hongzhi	1488–1505
11. Zhu Houzhao	Zhengde	1506–1521
12. Zhu Houcong	Jiajing	1522–1566
13. Zhu Zaihou	Longqing	1567–1572
14. Zhu Yijun	Wanli	1573–1620
15. Zhu Changle	Taichang	1620
16. Zhu Yujiao	Tianqi	1621–1627
17. Zhu Yujian	Chongzhen	1628–1644

Tables for Reference

Abbreviations Used in the Tables

CRB   Boxer, *The Great Ship from Amacon*

CSC   Boxer, *South China in the Sixteenth Century*

DP   Pantoja, *Advis du Reverend Père Iaques Pantoie de la Compagnie de Jésus*

HB   Wu Gang, *Huashan beishi*

HR   Hai Rui, "Xingge tiaoli," *Hai Rui ji* 1:38–145

IS   Inoue Susumu, *Chūgoku shuppan bunka shi*

LHL   Fan Lai, *Liangzhe haifang leikao xubian*

LW   Zhang Dai, *Langhuan wenji*

PHP   Purchas, *Purchas His Pilgrimes*, vol. 3

PYD   Pan Yunduan, *Yuhua tang riji*, as cited in Zhang Anqi, "Ming gaoben 'Yuhua tang riji' zhong de jingjishi ziliao yanjiu"

SB   Shen Bang, *Wanshu zaji*

SC   Shum Chum, "Mingdai fangke tushu zhi liutong yu jiage"

SF   *Songjiang fuzhi* [Gazetteer of Songjiang prefecture], 1630

TS   *Tianshui bingshan lu*, *Ming wuzong waiji* edition

WSQ   Wang Shiqiao, *Xiguan zhi*

WSX   *Wu Shangxian fenjia pu*, cited in Wu Renshu, *Youyou fangxiang*, 333

WX   *Wujiang xianzhi* [Gazetteer of Wujiang county], 1561

YMZ   Ye Mengzhu, *Yueshi bian*

TABLE 2.1. Posted Values of Furnishings for Officials' Residences, Zhejiang, 1562

		Unit price	Number of items assigned per residence		
			1st class	2nd class	3rd class
Furniture					
Bed rail	牀芭	0.01	4	8	2
Sedan chair bench	轎凳	0.02	2	1	0
Cot	小牀	0.04	2	2	2
Bed board	床板	0.06	3	2	2
Four-poster bed	四柱牀	0.08	2	2	0
Parasol	日傘	0.1	1	1	1
Umbrella	雨傘	0.13	1	1	1
Document table	案桌	0.15	1	1	1
Desk	官桌	0.25	6	4	4
Armchair	交椅	0.25	0	0	4
Armchair	交椅	0.275	0	4	0
Armchair	交椅	0.3	6	0	0
Summer bed	涼牀	0.5	1	1	1
Midseason bed	中牀	0.8	1	0	1
Small table skirt	小桌幃	0.8	1	1	1
Awning	涼傘	1.5	1	1	1
Winter bed	暖牀	1.8	1	1	0
Subtotal			9.58	7.54	6.2
Household furnishings					
Foot-warming stool	腳火凳	0.01	2	1	1
Folding ruler	摺尺	0.015	2	0	0
Plain bench	粗凳	0.02	2	1	0
Plain basin stand	粗面架	0.02	1	1	1
Plain clothes rack	粗衣架	0.03	1	1	1
Tub	桶盤	0.04	2	0	0
Shoe rack	靴架	0.05	1	1	1
Wooden chime with stand	木魚並架	0.06	1	1	0
Inkstone case (set)	硯匣	0.07	0	1	1
Inkstone case (set)	硯匣	0.08	1	0	0
Tea table	茶架	0.08	2	1	1
Decorated basin stand	花面架	0.08	1	1	0
Toilet	淨桶	0.1	1	1	1
Footbath	腳桶	0.1	1	1	1
Sitz bath	坐桶	0.12	1	1	1
Bath tub	浴桶	0.15	1	1	1
Small stove	方爐	0.16	1	1	0
Decorated clothes rack	花衣架	0.16	1	1	0
Court robe clotheshorse	執事架	0.5	1	0	0
Subtotal			1.93	1.21	0.73
Kitchenware					
Brush	刷帚	0.0025	4	1	0
Whisk	茗帚	0.005	2	1	1
Broom	掃帚	0.005	2	1	0
Earthenware bowl	鉢頭	0.005	2	1	0

		Unit price	Number of items assigned per residence		
			1st class	2nd class	3rd class
Nightsoil pan	糞箕	0.005	2	1	0
Colander	笊籠	0.005	1	1	0
Fire tongs (pair)	火筋	0.01	1	1	0
Spatula	鍋擀	0.01	1	1	0
Basin	挽桶	0.01	1	1	0
Wooden candlesticks (pair)	木燭臺	0.015	2	1	0
Wok lid	鍋蓋	0.015	3	1	2
Wooden stir stick	木筋	0.015	2	1	0
Teaspoons (set)	茶匙	0.02	2	1	1
Hatchet	斧頭	0.03	1	1	1
Kitchen knife	廚刀	0.03	2	1	1
Pots (set)	插盆	0.03	1	1	0
Rice cooker	飯甑	0.04	0	0	0
Rice-washing pail	淘米桶	0.04	1	1	0
Water barrel	水缸	0.04	2	1	1
Kitchen table	廚桌	0.05	1	1	1
Copper ladle	銅勺	0.05	1	1	1
Fire rake	火鍬	0.05	1	1	0
Steamers (set)	蒸籠	0.06	1	1	0
Curing basket	焙籠	0.06	1	1	0
Bamboo chair	竹椅	0.06	24	4	0
Water buckets (set)	水桶	0.08	1	1	1
Wok	鍋	0.1	3	1	2
Hibachi	火盆	0.1	1	1	0
Subtotal			2.67	1.085	0.535
Pewterware	錫器				
Basin	盆		1	1	1
Wine pot	酒壺		1	1	1
Pair of candlesticks	燭臺		2	1	1
Wine chafer	酒鏇		1	1	1
Teapot	茶壺		1	1	1
Inkstone case (set)	硯匣		1	1	1
Chamber pot	夜壺		1	1	1
Weight in catties			26.375	18.675	14.375
Value in taels			1.85	1.3	1
Porcelain	磁器				
Teacups	茶鍾		12	12	12
Serving plates	大白盤		10	10	10
Soup bowls	湯碗		10	10	10
Wine cups	酒盞		10	10	10
Saucers	白碟器		60	60	60
Value in taels			0.5	0.5	0.5
Subtotal			2.35	1.8	1.5
Total in taels			16.53	11.635	8.965

Source: Hai Rui, *Hai Rui ji*, 129–35.

TABLE 3.1. East India Company Annual Imports from Asia to Europe ca. 1620

Item	Consumed in Europe (lb.)	Cost in Aleppo		Cost in Asia		Price in London formerly		Price in London latterly	
		Unit cost (s)	Total cost (£)	Unit cost (s)	Total cost (£)	Unit cost (s)	Revenue (£)	Unit cost (s)	Revenue (£)
Pepper	6,000,000	2	600,000	0/2½	62,500	3/6	70,000	1/8	33,333
Cloves	450,000	4/9	160,875/10	0/9	16,875	8	16,000	0/6	12,000
Mace	150,000	4/9	335,626	0/8	5,000	9	9,000	6	6,000
Nutmeg	400,000	2/4	6,666/13/4	0/4	6,666/13/4	4/6	36,000	2/6	20,000
Indigo	350,000	4/4	75,833/6/8	1/2	20,416/12/4	7	52,500	5	37,000
Persian silk	1,000,000	12	600,000	0/8	400,000				
Total (£)			1,465,001/10		511,458/5/8		183,500		108,333/6/8
Silver (kg)			175,800		61,375		22,020		13,000
Value in taels			4,713,142		1,645,442		590,349		348,525

Note: 12 pence = 1 shilling (s); 20 shillings = 1 pound (£).

Source: Mun, Discourse of Trade, 268–69, 291–92.

TABLE 3.2. Some Prices in Manila ca. 1575 Compared to Domestic Prices

	Manila price		Comparative price in China			
	Unit	Price	Price	Place	Date	Source
Food						
Flour	peck	0.01	0.012	Beijing	1590	SB 151
Pepper	catty	0.05	0.065	Zhangzhou	1608	PHP 516
Sugar	peck	0.014	0.03	Beijing	1590	SB 122
Animals						
Buffalo calf	each	1.46	2	Shanghai	1596	PYD 299
Buffalo	each	2	8	Beijing	1590	SB 132
Footwear						
Shoes	pair	0.05	0.045	Guangzhou	1556	CSC 124
Blue satin shoes	pair	0.3	0.36	Guangzhou	1556	CSC 124
Silk						
Twisted silk	tael	0.115	0.08	Guangzhou	1600	CRB 179
Loose silk	tael	0.64	0.32	Wujiang	1561	WX 1561
Ceramics						
Bowl	each	0.01	0.01	Beijing	1577	SB 141
Small bowl	each	0.008	0.005	Beijing	1577	SB 141
Bowl	each	0.0163	0.01	Beijing	1577	SB 141
Rough porcelain	each	0.011	0.005	Shanghai	1628	YMZ 164
Small plate	each	0.0167	0.007	Beijing	1577	SB 141
Blue-and-white plates	each	0.0167	0.005	Chun'an	1562	HR 131
Furniture						
Big chest	each	1.3	1.2	Beijing	1590	SB 135
Chest	each	0.7	0.6	Beijing	1590	SB 135
Painted writing desk	each	0.4	0.3	Beijing	1590	SB 132
Desk	each	0.4	0.25	Jiangxi	1562	TS 162
Black writing desk	each	0.4	0.25	Chun'an	1562	HR 129
Table	each	0.8	0.4	Beijing	1590	SB 148
Writing seat	each	0.4	0.2	Jiangxi	1562	TS 162
Wooden bed	each	1.6	0.8	Chun'an	1562	HR 129
Ink						
Dried ink	stick	0.015	0.01	Chun'an	1562	HR 83
People						
"Black slave"	person	0.2	1	Shanghai	1593	PYD 290

Note: Manila prices are from Archivo General de Indias (ES.41091.AGI/16/Contaduria 1195).

TABLE 3.3. Some Offshore Prices Recorded by John Saris Compared to Domestic Prices

		Offshore prices			Domestic prices			
	Unit	Bantam	Borneo	Japan	Price	Place	Date	Source
		1608	1608	1614				
Textiles								
Satin	foot	0.168		0.34	0.035	Nanjing	1606	JFZ 49.74a
Damask	foot	0.126	0.056		0.04	Beijing	1572	SB 136
Gauze	foot			0.224	0.121	Songjiang	1630	SF 15.5a
Velvet	foot	0.232	0.14	0.442				
Raw silk	picul	0.593	0.593	0.433				
Food								
China-root (simlax)	catty			0.04	0.012	Canton	1600	CRB 180
White sugar	catty	0.03		0.05	0.032	Beijing	1590	SB 122
Honey	catty			0.6	0.032	Beijing	1590	SB 122
Pepper	catty	0.038		0.1	0.065	Zhangzhou	1608	PHP 516
Dyes and fragrances								
Sappanwood	catty			0.26	0.1	Beijing	1590	SB 133
Musk	catty	16.3	5.19	15	8	Canton	1600	CRB 180
Metal								
Lead	catty	0.044		0.088	0.033	Hangzhou	1604	LHL 6.66b
Iron	catty	0.044		0.04	0.048	Beijing	1590	SB 122
Copper	catty			0.1	0.07	Canton	1600	CRB 184
Mercury	catty			0.4	0.4	Canton	1600	CRB 180
Tin	picul		0.111	0.35				
Other materials								
Candle wax	catty		0.111	0.25	0.15	Beijing	1590	SB 129
Sandalwood	catty		0.296	0.1	0.4	Songjiang	1620	YMZ 161
Vermillion	catty			0.6	0.4	Beijing	1590	SB 127
Bezoar	tael		3.7		6	Beijing	1590	SB 151

Sources for Saris's prices: Purchas, *Purchas His Pilgrimes*, 3:506–18; Saris, *Voyage of John Saris*, 204–6.

TABLE 5.1. Percentage Change of Some Prices on the Valuation Lists of 1368 and 1451

Item		Unit	Price in paper wen, 1368	Duty in paper wen, 1451	Price: duty × 16.7, 1451	Percent change from 1368 to 1451 prices
Lead	黑鉛	catty	3,000	100	1,670	−44
Mid-length lining paper	中夾紙	100	10,000	340	5,678	−43
Twill damask	綾	bolt	160,000	6,700	111,900	−30
Felt hat	氈帽	each	4,000	170	2,839	−29
Millstones	大碌	pair	30,000	670	22,378	−25
Beeswax	黃蠟	catty	10,000	670	11,190	11
Pepper	胡椒	catty	8,000	670	11,190	39
Open-weave gauze	紗	bolt	80,000	6,700	111,900	39
Tin	錫	catty	4,000	340	5,678	42
Tempered iron	鐵	catty	1,000	100	1,670	67
Woven reed mat	蘆席	each	1,000	100	1,670	67
Three-shuttle cloth	三梭布	bolt	40,000	4,000	66,800	67
Plums	楊梅	catty	1,000	100	1,670	67
Iron wok	鐵鍋	each	8,000	1,000	16,700	109
Oxhide	牛皮	each	24,000	4,000	66,800	178
Copper	生錢	catty	4,000	670	11,190	180
Deerskin	鹿皮	each	20,000	3,400	56,780	184

Source: Da Ming huidian, 5.40a–43a, 179.2a–13b.

TABLE 5.2. Percentage Change of Some Prices between 1368 and 1562

Item		Unit	Huidian valuation, 1368	Yan Song valuation, 1562	Hai Rui valuation, 1562	Percent change from 1368 to 1562 prices
Twill damask	綾	bolt	2	1.2		−40
Armchair	交椅	each	0.3	0.2		−33
Beeswax	白蠟	catty	0.125	0.11		−12
Armchair	交椅	each	0.3		0.3	0
Iron wok	鐵鍋	each	0.1		0.1	0
Knife	小刀	each	0.025		0.03	0
Table	桌	each	0.125		0.15	20
Tin	錫	catty	0.05	0.06		20
Refined copper	熟銅	catty	0.05	0.076		52
Stool	杌	each	0.025	0.05		100
Table	桌	each	0.125	0.25		100
Women's sedan chair	女轎	each	1	2		100
Silk gauze	紗	bolt	1	2		100
Turban	包頭	each	0.0125	0.03		140

Sources: Da Ming huidian, 179.2a–13b; Tianshui bingshan lu, 157–64; Hai Rui ji, 129–30.

TABLE 5.3. Changes in Real Prices between 1368 and 1590

Item		Unit	Price, 1368		Price, 1590		Percent change in indexed prices
			Converted to silver taels	Indexed to a peck of rice	Price in silver taels	Indexed to a peck of rice	
Pepper	胡椒	catty	1	320	0.007	14	−96
Small pongee	小絹	bolt	0.25	800	0.026	52	−94
Linen	麻布	bolt	1	3,200	0.105	210	−93
Large iron wok	大鐵鍋	each	1	3,205	0.17	340	−89
Porcelain saucer	瓷碟碗	each	0.025	80	0.005	10	−88
Ink	墨	catty	1	3,200	0.3	600	−81
Coarse ramie cloth	粗苧布	bolt	0.275	880	0.1	200	−77
Goose	鵝	each	1	320	0.05	100	−69
Round straw hat	綾草帽	each	1	320	0.05	100	−69
Peck measure	斗	each	0.025	80	0.015	30	−63
Duck	鴨	each	0.05	160	0.03	60	−63
Lead	黑鉛	catty	0.0375	120	0.025	50	−58
Bean-fiber cloth	葛布	bolt	0.25	800	0.18	360	−55
Stool	凳	each	0.05	160	0.04	80	−50
Ebony chopsticks	烏木筯	pair	0.005	16	0.004	8	−50
Walnuts, hazelnuts	核桃榛子	catty	0.0125	40	0.01	20	−50
Dried dates, chestnuts	棗栗	catty	0.0125	40	0.01	20	−50
Chicken, pheasant	雞野雞	each	0.0375	120	0.0034	68	−43
Dog	犬	each	0.125	400	0.12	240	−40
Felt	毯段	piece	0.6	2,000	0.6	1,200	−40
Armchair	交椅	each	0.3	960	0.3	600	−38
Silver	銀	tael	1	3,200	1	2,000	−38
Gold	金	tael	5	16,000	5.4	10,800	−33

Damask	紗	bolt	1	3,200	1.1	2,200	−31
Meat	肉	catty	0.0125	40	0.015	30	−25
Seafood	魚鰲蝦蟹	catty	0.0125	40	0.015	30	−25
Water chestnuts, gorgons	菱茨	catty	0.0125	40	0.015	30	−25
Iron hoe	鐵鋤	each	0.025	80	0.03	60	−25
Flour	麵	catty	0.0062	20	0.008	16	−20
Copper	銅	catty	0.05	160	0.07	140	−13
Coarse cotton cloth	粗綿布	bolt	0.125	400	0.18	360	−10
Rice	米	peck	0.0312	100	0.05	100	0
Wheat	小麥	peck	0.025	80	0.04	80	0
Barley	大麥	peck	0.013	40	0.025	50	25
Peaches, pears	桃梨	100	0.025	80	0.05	100	25
Brush pens	筆	each	0.0025	8	0.005	10	25
Small knife	小刀	each	0.025	80	0.05	100	25
Poster paper	榜紙	100	0.5	1,600	1	2,000	25
Tin	錫	catty	0.05	160	0.1	200	25
Salt	鹽	catty	0.0031	10	0.0067	13	30
Beans of all sorts	黃黑綠豌豆	peck	0.0225	72	0.05	100	39
Bamboo chopsticks	竹筯	pair	0.0062	20	0.015	30	50
Apricots	杏	100	0.0125	40	0.03	60	50
Sesame oil	香油	catty	0.0125	40	0.03	60	50
Honey, sugar	蜂蜜沙糖	catty	0.0125	40	0.032	64	60
Ox	黃牛	each	3.125	10,000	8	16,000	60
Large screen	大屏風	each	0.3	960	0.8	1600	67
Cinnabar	硃砂	tael	0.05	160	0.15	300	88
Tea	茶	catty	0.0125	40	0.04	80	100
Wine, vinegar	酒酢	bottle	0.0125	40	0.04	80	100
Grapes	葡萄	catty	0.0125	40	0.04	80	100

(Continued on next page)

TABLE 5.3. (continued)

Item		Unit	Price, 1368		Price, 1590		Percent change in indexed prices
			Converted to silver taels	Indexed to a peck of rice	Price in silver taels	Indexed to a peck of rice	
Reed mat	廬席	set	0.0125	40	0.04	80	100
Sulphur	硫黄	catty	0.0125	40	0.04	80	100
Large wooden bucket	大木桶	each	0.0625	200	0.2	400	100
Table	卓	each	0.125	400	0.4	800	100
Drum	鼓	each	0.0625	200	0.2	400	100
Charcoal	灰炭	catty	0.0012	4	0.004	8	100
Memorial paper	奏本紙	100	0.2	640	0.7	1,400	119
Cayenne pepper	花椒	catty	0.0125	40	0.048	96	140
Iron	鐵	catty	0.0125	40	0.048	96	140
Board	木板	155 cm	0.05	160	0.2	400	150
Bamboo hat	笠	each	0.0125	40	0.05	100	150
Iron chain	鐵索	each	0.0125	40	0.06	120	200
Brick	磚	100	0.2	640	1.055	2,110	230
Sappanwood	蘇木	catty	0.0375	120	0.24	480	300
Notebook paper	手本紙	100	0.0875	280	0.6	1,200	329
Firewood	大柴	catty	0.001	3	0.007	14	367
Rabbit	兔	each	0.05	160	0.375	750	369
Log	木	3 m	0.075	240	0.6	1,200	400
Large colored letter paper	各色大箋紙	100	0.25	800	2	4,000	400
Brocade	錦	bolt	0.1	320	0.8	1,600	400
Mid-length lining paper	中箋紙	100	0.125	400	1.5	3,000	650

Sources: Da Ming huidian, 179.2a–13b; Shen Bang, Wanshu zaji, 121, 129–41, 145–48, 151, 170.

TABLE 5.4. Prices of Some Commodities in the Ledger of Cheng's Dyeworks, 1593–1604

			Quantity	Unit	Value in taels	Price per unit	Year
Indigo	Local indigo	土靛	10,000	catty 斤	100	0.01	1594
			16,300		160	0.01	1597
			2,000		14.4	0.0072	1598
	Rizhang indigo	日张靛	39,900	catty 斤	610.4	0.0153	1597
			38,200		687.6	0.018	1598
			30,750		399.7	0.013	1600
			34,100		405.7	0.0119	1601
	Courtyard indigo	园靛	7,200	catty 斤	93.6	0.013	1600
			12,000		131	0.0109	1601
			5,000		55	0.011	1603
Cloth	Dyed (blue) cloth	青布	13,800	bolt 疋	3,030	0.22	1593
			12,500		2,680	0.215	1594
			2,928		670.7	0.229	1597
			164		32.1	0.196	1597
			8,057		1798	0.223	1598
			14,219		3,185.5	0.224	1599
			10,138		2,228	0.222	1600
			4,002		875.9	0.219	1601
			8,214		1,800	0.219	1602
			6,770		1,538	0.227	1603
			7,833		1,793.7	0.229	1604
	Undyed cloth	白布	4,168	bolt 疋	625	0.15	1594
			5,073		811.7	0.16	1597
			680		98.2	0.144	1597
			12,172		1,862.3	0.153	1598
			7,296		1,079.8	0.148	1599
			5,100		790	0.155	1600
			8,209		1,238	0.151	1601
			9,464		1,701	0.18	1602
			10,016		1,596.7	0.159	1603
			10,885		1,720	0.158	1604
	Gangshang cloth	缸上白布	387	bolt 疋	58	0.15	1594
			625		95.6	0.153	1598
			678		99.5	0.147	1599
			583		90	0.154	1600
			682		102.84	0.151	1601

(Continued on next page)

TABLE 5.4. (continued)

		Quantity	Unit	Value in taels	Price per unit	Year
		585		86.6	0.148	1602
		452		72	0.159	1603
		592		94.7	0.16	1604
Bean fiber cloth	葛布	450	bolt 疋	67.5	0.15	1593
		181		28	0.155	1597
		339		51	0.15	1598
		238		35.7	0.15	1599
		130		19.5	0.15	1600
		200		32	0.16	1601
		37		5.7	0.154	1602
		71		10.6	0.149	1603
		110		18	0.164	1604
Cloth dyed elsewhere	客染布	91	bolt 疋	5	0.055	1598
		60		3.5	0.058	1599
		57		3.7	0.065	1600
		50		3	0.06	1601
		71		3.5	0.049	1603
		121		6.5	0.054	1604
Privately dyed book cloth	私染書布	11	bolt 疋	2.2	0.2	1600
		7		1.4	0.2	1601
Distillery dregs	糟	200	jar 埕	25	0.125	1594
		220		22	0.1	1598
		214		21	0.098	1599
		230		23	0.1	1600
		190		18	0.095	1601
		155		15.5	0.1	1602
		153		15.3	0.1	1603
Firewood	柴	700	catty 斤	10	0.014	1598
		750		10	0.013	·1599
		1,000		11	0.011	1600
		700		10.4	0.015	1601
		1,700		19	0.011	1603
Lime (calcium hydroxide)	灰	4	container 房	12	3	1598
		4		15	3.33	1599
		4		15	3.33	1600
		4		12	3	1601

(Continued on next page)

		Quantity	Unit	Value in taels	Price per unit	Year
		3		10	3.33	1602
		2		6	3	1603
Rice	米	60	hectoliter 石	36	0.6	1593
		16		11	0.687	1598
		71		50	0.704	1599
		10		6	0.6	1600
		14		7.7	0.55	1601
		20		15	0.75	1602
		2		1.5	0.75	1604
Rush wrappings	蒲包	1,220	each 个	3	0.0025	1600
		1,800		4.5	0.0025	1603
		2,200		5	0.0023	1604
"Cloth heads"	布头	175	catty 斤	17.5	0.1	1597
		500		50	0.1	1598
		190		19	0.1	1599
		514		50	0.097	1600
		530		52	0.098	1601
		19		2.2	0.116	1603
		32		3.8	0.119	1604
"Shaved skins"	希皮	2,400	each 个	4	0.00167	1598
		2,800		4.7	0.00168	1600
		1,400		2	0.00143	1601
		1,700		2.7	0.00159	1602
		1,550		2.63	0.0017	1603

Source: Chengshi randian chasuan zhangbu, 8, 9, 11, 19, 20, 28, 29, 36, 44, 54, 55, 65, 71, 77, 78.

NOTES

Preface. My Brief Life as a Price Historian

1. Le Roy Ladurie, *Histoire humaine et comparée du climat*, 17–29. On the physical proxies for the Little Ice Age, see Mann et al., "Global Signatures and Dynamical Origins of the Little Ice Age and Medieval Climate Anomaly"; Campbell, *Great Transition*, 335–44; Degroot, *Frigid Golden Age*, 2–9, 31–41. For a recent summary of tree-ring data for this period, see Wilson et al., "Last Millennium Northern Hemisphere Summer Temperatures." My research supports an early onset of the Little Ice Age in China; Brook, "Nine Sloughs," 30–44.

2. On the concept of the Great State, see Brook, "Great States."

3. That story is told in rich detail in Parsons, *Peasant Rebellions of the Late Ming*; and Wakeman, *Great Enterprise*.

4. On the emergence of climate history among historians of Europe, see Le Roy Ladurie, "Birth of Climate History."

5. Yang Lien-sheng, *Money and Credit in China*, 103.

6. Peng Xinwei, *Zhongguo huobi shi*, 442–69; Kaplan, *Monetary History of China*, 597–616.

7. Nakayama, "On the Fluctuation of the Price of Rice"; Nakayama, "Shindai zenki Kōnan no bukka dōkō." Other contributions to Qing price history at this time include Y. Wang, "Secular Trend of Prices during the Ch'ing Period"; Marks, "Rice Prices, Food Supply, and Market Structures"; and Will and Wong, *Nourish the People*.

8. Frank, *ReOrient*, ch. 2.

9. An energetic and informed critique of Frank's assumptions is Deng, "Miracle or Mirage."

Chapter 1. The Tale of Chen Qide

1. Chen Qide, "Zaihuang jishi" (A record of events during the disaster-driven famine), printed both in his collected writings, *Chuixun puyu*, 16a–20a, and in the local gazetteer *Tongxiang xianzhi* (1887), 20.*Xiangyi*.8a–10a. Parenthetical page references in this chapter are to the version in *Tongxiang xianzhi*.

2. Chen Qide, *Chuixun puyu*, 14b–15a.

3. For a useful overview of silver as a medium of exchange through the latter half of the Ming, see Kuroda, "What Can Prices Tell Us about 16th–18th Century China?"

4. Finlay, *Pilgrim Art*, 33, notes that the Malay word derived from the Tamil name for a small tin coin, *karshápana*, which in turn gave the Chinese their word *jiazhi*, "value" or "price." This etymology is unlikely, as the word *jiazhi* was already attested in the Han dynasty.

5. Paper: Peng Xinwei, *Zhongguo huobi shi*, 477n2; charcoal, chopsticks: Hai Rui, *Hai Rui ji*, 88, 130.

6. Li Fang, a senior member of the wealthiest family in Jiaxing, noticed a copper lying in the dirt outside his gate and walked past it. Later that day it struck him that there had been something odd about the coin. When he went back to get it, he discovered that someone else had picked it up. Li sent a servant to find the man, suspecting it was a "bad penny," a counterfeit coin cast from a debased alloy. When the servant found the man, Li offered to exchange it for a good copper of his own. See Tan Qian, *Zaolin zazu*, 593.

7. That ordinary Ming Chinese were concerned to count every copper was much repeated by foreign observers, e.g., Ch'oe Pu, *Record of Drifting across the Sea*, 157. On the copper coin in late Ming literature, see Shan, "Copper Cash in Chinese Short Stories," 230–35.

8. Girard, *Le voyage en Chine*, 125.

9. Chen Qide, *Chuixun puyu*, 6a.

10. E.g., Zhang Mou (1437–1522) associated price fairness with public justice when he criticized officials for refusing to allow grain to come in from neighboring areas to lower famine prices as supporting "petty self-advantage for the few" rather than acting "in the public interest (*gongping*) that is just for all"; Chen Zilong, *Huang Ming jingshi wenbian*, 95.14a.

11. Xu Guangqi, *Nongzheng quanshu*, 194.

12. Shanghai bowuguan tushu ziliao shi, *Shanghai beike ziliao*, 82–83. On the recognition that merchants had to work across a price gap, see Zhang Han, *Songchuang mengyu*, 80, translated in Brook, "Merchant Network in Sixteenth Century China," 187.

13. De Vries, *Price of Bread*, 9–10.

14. Muldrew, *Economy of Obligation*, 44–46, where he cites the quotations of Malynes and Scott.

15. E.g., when a county gazetteer records that the grain harvested from fields donated to the county school "was converted into silver at the current price," the intention is to stamp the sale with approval; *Yushan xianzhi* (1873), 4a.37b, quoting from an earlier Ming edition. So too, when a minister of revenue in 1524 proposed converting the grain rations for soldiers working on Grand Canal grain boats, he called for it to be done at the "market price" to legitimize the action; Xie Bin, *Nanjing hubu zhi*, 10.18a.

16. Xu Guangqi, *Xu Guangqi ji*, 459. As an earlier Ming gentleman advised his sons, the prosperity of the household depended on the ability to "apprise oneself of fluctuations in the price of grain"; translation in Ebrey, *Chinese Civilization and Society*, 198.

17. Yu Sen, *Huangzheng congshu*, 1.5b–6a. On the concept of market price in the Qing context, see Will, "Discussions about the Market-Place," 328–29.

18. Xie Bin, *Nanjing hubu zhi*, 17.4b, quoting from the official administrative manual, *Zhusi zhizhang* (Duties of office for all administrative posts). European monarchs too required local officials to send them reports of local grain prices; see De Vries, *Price of Bread*, 22.

19. Liu Ruoyu, *Zhuozhong zhi*, 101.

20. In 1570, the checking of prices that had been required in the first and seventh lunar months was moved to the fifth and eighth months (June and September) in response to merchants' complaints that prices in the first and seventh months were below actual half-year averages; Li Jiannong, "Mingdai de yige guanding wujia biao yu buhuan zhibi," 257.

21. For a selection of imperial edicts requiring officials to respect current prices, see *Da Ming huidian*, 37.31a–33b.

22. Xie Bin, *Nanjing hubu zhi*, 17.5b.

23. Koo Bumjin, *Imun yŏkchu*, 151.

24. Quoted in Li Jiannong, "Mingdai de yige guanding wujia biao yu buhuan zhibi," 257.

25. Zhang Kentang reports doing this in a case in which he judged the seller of fifty *mu* of land to have charged a buyer far in excess of the price he originally paid for it, ordering the seller to reduce his price by 0.3 taels per mu; *Xunci*, 6.27b.

26. Ye Chunji, *Huian zhengshu*, 11.11b.

27. Zhang Han, *Songchuang mengyu*, 143.

28. Quoted in Kawakatsu, *Min-Shin Kōnan nōgyō keizai shi kenkyū*, 209.

29. While the upholding of fair prices is consistent with the Confucian principle that the state should intervene on behalf of the people's welfare, this concern for prices is hardly unique. On the basis of a very different body of philosophy, canon law, and administrative practice, local authorities in England actively sought to regulate commodity prices throughout the period we would describe as the first half of the Ming dynasty, only gradually thereafter giving a freer hand to the market; see Britnell, "Price-Setting in English Borough Markets," 15.

30. Huang Zhangjian, "Ming Hongwu Yongle chao de bangwen junling" (1977), reprinted in his *Ming-Qing shi yanjiu conggao*, 275, 282.

31. Ming rules about fair pricing largely derived from Yuan law. The Yuan government insisted that officials pay what a censor in 1310 called "the real price in the streets and markets." This requirement was already in place thanks to a 1283 directive that officials send runners into the markets every month to check the price of "commodities in the streets and markets" and report them upward; *Da Yuan shengzheng guochao dianzhang*, 26.3a–4a, 6a. According to an imperial edict of 1341, if an official forced a price down, he should be punished under bribery rules, which meant that the greater the discrepancy, the heavier the penalty; Han'gukhak chungang yŏn'guwŏn, ed., *Chijŏng chogyŏk*, 97. Harmonious purchasing is not as visible in Ming documents until 1437, when central regulations allowed local officials and merchants to meet to negotiate the prices of official purchases so as to ensure that goods the officials needed to buy did not slip off the market for want of adequate price incentives; Su Gengsheng, "Mingchu de shangzheng yu shangshui," 436.

32. Officials were also supposed to respect local prices when assessing the value of land for tax purposes; Gu Yanwu, *Tianxia junguo libing shu*, 8.77b, from a text dated 1572.

33. Le Goff, *Money and the Middle Ages*, 144–45.

34. *Shexian zhi* (1609), 6.12a, quoted in Brook, *Confusions of Pleasure*, 238.

35. Zhang Yi, *Yuguang jianqi ji*, 502, dated to the reinstatement of Gao Gong as chief grand secretary.

36. Li Le, *Jianwen zaji*, 11.42b–44b.

37. Hamilton, "Use and Misuse of Price History," 47.

38. Shen Bang, *Wanshu zaji*, 141.

39. Hai Rui, *Hai Rui ji*, 130.

40. Rusk, "Value and Validity," 471.

41. Hamilton, "Use and Misuse of Price History," 48.

42. Cartier, "Note sur l'histoire des prix," 876.

43. Beveridge, *Wages and Prices in England*, xxvi.

44. Klein and Engerman, "Methods and Meanings in Price History," 9.

45. Gibson and Smout, *Prices, Food and Wages in Scotland*, 14.

46. Braudel, *Structures of Everyday Life*, 27.

47. On the effect of the difference between print and manuscript sources on writing the histories of China and Europe, see Brook, "Native Identity under Alien Rule," 237–39.

48. On the application of food energy to agriculture in preindustrial economies, see Muldrew, *Food, Energy and the Creation of Industriousness*, chs. 1–2.

49. Reddy, *Money and Liberty in Modern Europe*, 63–73.

50. In a review of fourteen Chinese reconstructions of long-term shifts in temperature, Ge Quansheng and his colleagues found that the variations in regional findings were pronounced when the data were resolved at the decadal level, but effectively disappeared when resolved to units of thirty years; see "Coherence of Climatic Reconstruction."

51. On the values and limits of the documentary approach to climate history, see Alexandre, *Le climat en Europe*, 9–42. Alexandre draws from a remarkable ensemble of 2,390 texts selected as reliable from a candidate corpus of over 3,500 texts, plus another 440 notices extracted from other documents such as account books.

52. Bauernfeind and Woitek, "Influence of Climatic Change," 304.

53. E.g., Campbell, *Great Transition*, 45, 57, 341.

54. Campbell, *Great Transition*, 345.

55. Y. Y. Kueh pursues a conceptually related project when he proposes that areas struck by grain shortage can be taken as proxies for disturbed weather, especially in the short term, in his *Agricultural Instability in China*, 286–300.

56. Xie Zhaozhe, *Wu zazu*, 31. Agrarian prognostications in the early Qing are discussed in Agøy, "Weather Prognostication in Late Imperial China as Presented in Local Gazetteers."

Chapter 2. Halcyon Days? The Wanli Price Regime

1. On the life of the young Wanli emperor, see the first four chapters of R. Huang, *1587, a Year of No Significance*.

2. *Ming Shenzong shilu* (Veritable records of the Wanli reign), 271.1a (year 22, month 3, *simao*). This passage, with two minor variations, is appended to Wang Xijue, "Quanqing zhenji shu" (Memorial strenuously urging aid), reprinted in Chen Zilong, *Huang Ming jingshi wenbian*, 395.7a–b. For the full context of this conversation, see Brook, "Telling Famine Stories."

3. *Ming shenzong shilu*, 271.1a. References in the *Veritable Records* suggest that Wanli's version of events in his reply to Wang Xijue was a collage of moments. For example, his comment about having viewed the *Album of the Famished* was made on 5 April (*jiazi*) (270.4a), whereas he received Wang's proposal to divert official salaries on 20 April (*simao*) (271.1a).

4. *Ming shenzong shilu*, 270.4a (Wanli 22, month 2, *jiazi*).

5. *Jimin tushuo* survives in a 1748 copy of a 1658 printing in the Henan Provincial Museum in Zhengzhou. I am grateful to Roger Des Forges for sharing his copy with me.

6. Contributions were made by the empress dowager (*Ming shenzong shilu*, 271.4a); Wanli's son by Zheng Guifei, Prince Fu (Zhu Changxun, 1581–1641) (273.2a); and Prince Shen (Zhu Chengyao) (273.4b), among others. Exact figures for the total relief are difficult to ascertain. Partial data can be found in Yang Dongming, *Jimin tushuo*, 38b; Wang Xijue, "Quanqing zhenji shu," in Chen Zilong, *Huang Ming jingshi wenbian*, 395.7a; *Ming shenzong shilu*, 271.1a, 9a.

7. Lu Zengyu, *Kangji lu*, 3a.66a. On the use of market forces to relieve the 1594 famine, see Yim, "Famine Relief Statistics," 5–7.

8. From the biography of Yang Dongming in his county gazetteer, *Yucheng xianzhi*, 1895, 6a.12a.

9. Libbrecht, *Chinese Mathematics in the Thirteenth Century*, 431. The solution to another problem in the same section of the textbook results in the price of rice being over twice the price of wheat, which is unlikely but not quite so unrealistic.

10. Martzloff, *History of Chinese Mathematics*, 47.

11. Sun and Sun, *Chinese Technology in the Seventeenth Century*, xi, with minor revisions. Although not published until 1637, the book captures much of the mood and many of the changes that prevailed during the Wanli era, On the response of Song's cosmology to "the growing presence of material things and practical matters" in the life and work of people in the late Ming, see Schäfer, *Crafting of the Ten Thousand Things*, 129.

12. Li Rihua, *Weishui xuan riji*, 103, quoted in Brook, "Something New," 369.

13. Pantoja, *Advis du Reverend Père Iaques Pantoie*, 111–12.

14. Matteo Ricci makes the same observation; see Gallagher, *China in the Sixteenth Century*, 12.

15. Armchairs ranged from as low as 0.2 taels (*Tianshui bingshan lu*, 162) to 0.4 taels (Shen Bang, *Wanshi zaji*, 148), and even as high as 0.5 taels (Wu Renshu, *Youyou fangxiang*, 334).

16. As Tina Lu has noted, the seventeenth-century market for luxury items was segmented in such a way that price differentials served to distinguish goods not only by quality but also by the social status of appropriate buyers; "Politics of Li Yu's *Xianqing ouji*," 495.

17. Crossbow: Fan Lai, *Liangzhe haifang leikao xubian*, 6.65a; turban: *Tianshui bingshan lu*, 161; Ding-ware dish: Zhang Anqi, "Ming gaoben 'Yuhua tang riji' zhong de jingjishi ziliao yanjiu," 306; ten geese: Shen Bang, *Wanshi zaji*, 170.

18. County magistrates were required to submit summary accounts of the funds in their treasuries regularly to circuit officials for inspection; see Li Le, *Jianwen zaji*, 3.111b.

19. That set of rules survives in the collection of Hai's writings on Chun'an completed in 1562, "Xingge tiaoli" (Regulations initiated and suspended), in his "Chun'an xian zhengshi" (Administrative affairs of Chun'an county), reprinted in *Hai Rui ji*, 38–145. For an account of Hai's fiscal reforms, albeit without reference to prices, see Cartier, *Une réforme locale*, 56–84.

20. Hai Rui, *Hai Rui ji*, 72.

21. Hai Rui, *Hai Rui ji*, 38.

22. Zhu Fengji's widely read magistrate's handbook *Mumin xinjian* (Mirror of the mind for shepherding the people) stresses the importance of buying at market prices, advising the magistrate to keep a ledger for public inspection to show what he paid for every item (1.6a).

23. Hai Rui, *Hai Rui ji*: "Rites," 81–89; "War," 105; "Works," 129–35. The introduction to the third of these lists is translated in Cartier, *Une réforme locale*, 145–46.

24. Hai Rui, *Hai Rui ji*, 128.

25. Wu Chucai, postface to Shen Bang, *Wanshu zaji*, 301.

26. Shen Bang, *Wanshu zaji*, 171. On the principle of *liang ru wei chu* (measuring revenue to determine expenditure), see Grass, "Revenue as a Measure for Expenditure," 96.

27. Shen Bang, *Wanshu zaji*, 121–74.

28. Not all the prices in the three tables strictly speaking fall within the Wanli era. Hai Rui's prices predate the Wanli era by a decade, and a few prices taken from other sources fall beyond the end of the era. I include them to broaden the range of items for inspection.

29. Shen Bang, *Wanshu zaji*, p. 170 (4 cents), p. 147 (5 cents), pp. 124–28 (6.4 cents), p. 170 (20 cents).

30. We learn this from stele inscriptions that the magistrate of Huayin county in Shaanxi erected in 1615 on Hua Mountain, one of the five sacred mountains of China. Concerned to control expenses, he posted that his office would pay 1 cent per catty for the meat sacrificed at state shrines to the spirit of Hua Mountain, with the exception of sacrificial deer, for which he paid 2.23 taels each; Wu Gang, *Huashan beishi*, 305, 306.

31. Pantoja, *Advis du Reverend Père Iaques Pantoie*, 112. I should note that I have found no other reference to fish at that price.

32. On the introduction of tobacco, see Brook, *Vermeer's Hat*, 120–23, 134–36.

33. Shen Bang, *Wanshu zaji*, p. 134 (2 cents), p. 146 (3 cents), p. 150 (10 cents).

34. The two handbooks furnish almost one hundred price reports for a multitude of types of paper. Shen's paper prices appear in Shen Bang, *Wanshu zaji*, 121–30, 137, 139, 145–46. Kai-wing Chow notes a few of Shen's prices in *Publishing, Culture, and Power in Early Modern China*, 35. I have not found an English equivalent for the cut. The English counted sheets of paper in 24s (a quire, often rounded up to 25) and 500s (a ream), but not in the 100s.

35. Shen Bang, *Wanshu zaji*, 123, 125, 126, 128, 10; Ye Mengzhu, *Yueshi bian*, 159.

36. In an extended discussion of tea grown in his native province of Fujian, Xie Zhaozhe writes that highly prized Songluo tea, which was labor intensive to produce, was in such demand that it created a frenzy of competition among retailers, the effect of which was to drive down the price of Songluo to 100 coppers (roughly 14 cents) per catty, a price point that made it impossible for any tea dealer to make any money. "How can the cost of labor be requited at that price?" Xie asks. "If the price were just slightly higher, however, no one would sell it" for lack of buyers. "This is why the demand for Fujian tea has recently declined." See Xie Zhaozhe, *Wu zazu*, 213.

37. Cited in Bian, *Know Your Remedies*, 136–37.

38. Girard, *Le voyage en Chine*, 253.

39. Dai, "Economics of the Jiaxing Edition," 331–33. The rate was cut by half toward the end of the dynasty in a bid to sell in greater volume, but the price returned to 6 cents in 1647.

40. Clunas, *Superfluous Things*, 132, citing the research of Isobe Akira, *Saiyūki juyōshi no kenkyū* (A study of the reception history of *Journey to the West*). On the market distinction between wealthier and poorer readers, see Hegel, "Niche Marketing for Late Imperial Fiction."

41. Paethe and Schäfer, "Books for Sustenance and Life," 19–20, 46.

42. Lu Wenheng, *Se'an suibi*, 2.13a.

43. Zhang Anqi, "Ming gaoben 'Yuhua tang riji' zhong de jingjishi ziliao yanjiu," 289.

44. When Wang Yangming in the 1510s set a fine of one tael for those who failed to attend the obligatory monthly meeting of their community compacts, he was imposing a severe penalty; Wang Yangming, *Instructions for Practical Living*, 300.

45. *Tianshui bingshan lu*, 160.

46. Feng Menglong, *Stories Old and New*, 462–64.

47. *Qingyuan zhilüe* (1669), 7.6a.

48. Girard, *Le voyage en Chine*, 239–55. The shipwreck is described more fully in Brook, *Vermeer's Hat*, 87–99, 109–13.

49. Girard, *Le voyage en Chine*, 239.

50. This price is roughly confirmed in S. Dyer, *Grammatical Analysis of the "Lao Ch'i-ta,"* 266; and Zhang Kentang, *Xunci*, 6.13b.

51. Girard, *Le voyage en Chine*, 243.

52. Girard, *Le voyage en Chine*, 244.

53. Girard, *Le voyage en Chine*, 246.

54. Zhang Yi, *Yuguang jianqi ji*, 324.

55. *Daming fuzhi* (1506), 1.12a.

56. Zhang Kentang, *Xunci*, 5.19a. I am grateful to Yonglin Jiang for alerting me to this source. In his "Defending the Dynastic Order at the Local Level," Jiang translates the title as "Court Verdicts That Touch the Heart." Will, *Handbooks and Anthologies for Officials in Imperial China*, 704, proposes "Plowing Words." In my view, neither best captures the author's intention.

57. Zhang Kentang, *Xunci*, 5.19a. It is not clear to me that the Ming Code imposes decapitation for this crime.

58. Zhang Kentang, *Xunci*, 1.14a, 5.14b, 3.26a.

59. In order of reference, Zhang Kentang, *Xunci*, 1.16a, 6.18b, 6.2b, 5.25b–26a, 1.16a, 24a, 3.2a–b, 6.22b.

60. Zhang Kentang, *Xunci*, 1.16a.

61. Niida Noboru, *Chūgoku hōseishi kenkyū*, 268.

62. Zhang Kentang, *Xunci*, 1.18b, 1.22b, 2.6a, 4.14, 2.8a.

63. Edvinsson and Söderberg, "Evolution of Swedish Consumer Prices," 415.

64. Allen, *British Industrial Revolution*, 35–37.

65. Studies of nineteenth-century men from south China suggest an average height of just over 163 cm; see Ward, "Stature, Migration and Human Welfare," 497.

66. Zhuang Yuanchen, *Manyan zhai cao*, cited in Chen Xuewen, *Huzhou fu chengzhen jingji shiliao leizuan*, 52–53.

67. Shen shi, *Bu nongshu*, 1.18a.

68. Zhang Lixiang, *Bu nongshu jiaoshi*, 142.

69. Li Le, *Jianwen zaji*, 7.15a. See also *Ming xuanzong shilu*, 45.8a (1428); Chen Zilong, *Huang Ming jingshi wenbian*, 481.25b; Tu Long, "Huangzheng kao," 181; Shen Bang, *Wanshu zaji*, 89.

70. Girard, *Le voyage en Chine*, 113, 324.

71. Lü Kun *Shizheng lu*, 2.52a.

72. Ge Yinliang, *Jinling fancha zhi*, 5.10b.

73. Wei Dazhong, *Wei Guoyuan xiansheng zipu*, in Huang Yu, *Bixue lu*, 20a.

74. Dean and Zheng, *Fujian zongjiao beiwen huibian: Quanzhou fu fence* 1:100.

75. For example, one prefectural gazetteer notes that wages for prefectural and county labor were in practice discounted to rates below what were published; *Songjiang fuzhi* (1630), 9.31a.

76. Chen Zilong, *Huang Ming jingshi wenbian*, 63.24a.

77. Xu Guangqi, "Gongcheng xinming jinchen jiqie shiyi shu," in Chen Zilong, *Huang Ming jingshi wenbian*, 488.25b; reprinted in Xu Guangqi, *Xu Guangqi ji*, 131; quoted in Brook, *Confusions of Pleasure*, 154. Liang Jiamian dates this memorial to 1619; *Xu Guangqi nianpu*, 124. Twenty-four coppers is given as the daily pay of soldiers manning the city wall in the defense of Yangzhou against the Qing armies in 1645, with the note that this was "not even enough for meals that satisfy hunger": Struve, *Voices from the Ming-Qing Cataclysm*, 12.

78. Shen Bang, *Wanshu zaji*, 130, 142, 144, 152. For the same rate in other contexts, see *Chuan zheng*, 39a, 40a; Huang Miantang, *Mingshi guanjian*, 369.

79. *Jingdezhen taoci shigao*, 105; Gerritsen, *City of Blue and White*, 180–81, with slight changes to her calculations.

80. Wu Yingji, *Liudu jianwen lu*, 13b.

81. Huang Xingzeng, *Can jing*, cited in Chen Xuewen, *Huzhou fu chengzhen jingji shiliao leizuan*, 59.

82. Qi paid these wages in silver, leaving it to his employees to exchange their silver for copper so that they could use it. The rates for officers were higher: 6 cents for platoon leader, 8 cents for a sergeant, and 10 cents for a captain; *Qi Biaojia ji*, 35, 123; *Qi Zhongmin gong riji*, 3.29, 4.8. Referring to military service in 1643, Tan Qian notes that special military salaries could go even higher, from 14 cents a day for muleteers to a full tael for senior officers, though these were danger-pay rates; *Zaolin zazu*, 115.

83. Dean and Zheng, *Fujian zongjiao beiwen huibian: Xinghua fu fence*, 103; see also *Putian xianzhi* (1879), 4.4b.

84. Girard, *Le voyage en Chine*, 243, converting from his figure of ¼ real.

85. Huang, *Taxation and Governmental Finance*, 120.

86. *Jianning fuzhi* (1541), 14.68a.

87. The need to align fiscal wages with real wages is attested in a county gazetteer of 1548 from Shandong, in which it is noted that the grain measurers working at the county granary used to be paid 2 taels, but that this wage had "now" been increased to 3 taels; *Laiwu xianzhi* (1548), 3.3a.

88. Night crier: Shen Bang, *Wanshu zaji*, 53; porter: *Songjiang fuzhi* (1630), 9.39b.

89. Scribe: *Qingliu xiaazhi* (1545), 2.35a; mounted messenger: *Linqu xianzhi* (1552), 1.44b.

90. Militia captain: *Zichuan xianzhi* (1546), 4.58a; station master: *Xiajin xianzhi* (1540), 2.24a.

91. *Ming chongzhen changbian*, 41.2b.

92. *Da Ming huidian*, 39.1b–7b. These official salaries in silver were heavily supplemented by "firewood and fuel" and other allowances; see Feng Mengzhen, *Kuaixue tang riji*, 72.

93. Huang, *Taxation and Governmental Finance*, 276. Contemporaries confirm this observation. In a diatribe against officials taking bribes, Xie Zhaozhe estimates that officials at ranks 5 to 7 could count on an informal annual income of 100 taels, and officials at rank 4 and above, double that figure; *Wu zazu*, 310.

94. On the difficulty of estimating standards of living in a society in which income is unevenly distributed, see Coatsworth, "Economic History and the History of Prices in Colonial Latin America," 27.

95. Poor man's bed: Hai Rui, *Hai Rui ji*, 129; rich man's bed: *Tianshui bingshan lu*, 160.

96. The poor man lived in a chapel at the foot of Pan Mountain in the 1590s on 2,000 copper cash a year, earned from selling apples from a huge tree outside his chapel; Tang Shisheng, "You Panshan ji" (A record of traveling to Pan Mountain), in *Panshan zhi buyi* (1696), 1.1b. The wealthy man is Zhuang Yuanchen, who complained that he spent 400 taels in 16 months preparing for the exams; Hamashima, "Minmatsu Kōnan kyōshin no gutaisō," 178.

97. Feng Qi, "Su guanchang shu" (Memorial criticizing the common practices of officials), quoted in Xu Hong, "Mingmo shehui fengqi de bianqian," 108. Feng complained further that these extravagant officials sometimes spent an appalling 2 to 3 taels on a single meal.

98. Girard, *Le voyage en Chine*, 249–50, 253.

99. Girard, *Le voyage en Chine*, 244.

100. Dudink, "Christianity in Late Ming China," 182.

101. Zhang Yi, *Yuguang jianqi ji*, 430.

102. Li Rihua, *Weishui xuan riji*, 246.

103. As Tina Lu has argued in her study of the writer Li Yu, Li "explains that rich people compete to buy antiques at high prices, sometimes justifying their behavior as a desire to commune with the ancients. But . . . antiques only make sense as an efficient means to store and transport vast quantities of wealth. Without wealth, the very notion of antique is irrelevant"; "Politics of Li Yu's *Xianqing ouji*," 498.

104. Quoted in Heijdra, "Mingdai wenwu dagu Wu Ting shilüe," 404.

105. Oertling, *Painting and Calligraphy*, 129.

106. Li Rihua, *Weishui xuan riji*, 30–32.

107. Yu Jianhua, *Zhongguo meishujia renming cidian*, quoted in Clunas, *Elegant Debts*, 123.

108. Kuo, "Huizhou Merchants as Art Patrons," 180.

109. My principal sources for art prices are Clunas, *Superfluous Things*, 179–80; *Da Ming huidian* 179.2a; Girard, *Le voyage en Chine*, 252; Li Le, *Jianwen zaji*, 3.33b, 10.35a; Li Rihua, *Weishui xuan riji*, 246, 401; Shen Defu, *Wanli yehuo bian*, 663; *Tianshui bingshan lu*, 159; Wu Renshu, *Youyou fangxiang*, 333; Yuan Zhongdao, *Youju feilu*, 248; Zhang Anqi, "Ming gaoben 'Yuhua tang riji' zhong de jingjishi ziliao yanjiu," 298–309; Zhang Dai, *Taoan mengyi*, 7.

110. Ye Kangning, *Fengya zhi hao*, 202–19. I have excluded the price of 2,000 taels that Xiang allegedly paid for a Tang copy of Wang Xizhi's calligraphy, which his descendants allegedly recouped when they sold it in 1619, as an unreliable outlier.

Chapter 3. Silver, Prices, and Maritime Trade

1. This interpretation was first proposed by William Atwell in 1977 in "Notes on Silver, Foreign Trade, and the Late Ming Economy," followed in 1982 by "International Bullion Flows and the Chinese Economy." Atwell's arguments were championed in the field of global history by, among others, Flynn and Giráldez in their "Born with a 'Silver Spoon.'" The effect of foreign silver on China's economy has been more critically examined by von Glahn in several publications, notably *Fountain of Fortune*, 113–41. For a spirited riposte to his critics, see Atwell, "Another Look at Silver Imports into China."

2. Chen Qide, *Chuixun puyu*, 15a. Chen's pleasure at not having been born outside China was not an insight of his own making but repeats the second of the five privileges of the literati that Lü Nan listed early in the sixteenth century; see Bol, *Localizing Learning*, 10.

3. One did not have to be from the south coast to get entangled in maritime trade, as the tale of the Suzhou silversmith Guan Fangzhou attests; Brook, *Troubled Empire*, 213–15.

4. Zhang Xie, *Dongxi yang kao*, 170.

5. Tan Qian, *Zaolin zazu*, 483–84.

6. Eight reals made a peso, is why the English called pesos "pieces of eight."

7. Von Glahn, *Fountain of Fortune*, 133–37.

8. In an insightful reassessment of the role of silver in international trade, Niv Horesh argues that China did not simply draw silver automatically by virtue of its economy's capacity to respond to demand, but that European traders adopted a conscious strategy of using silver to reallocate precious metals between regions so as to maximize their purchasing power. As he writes, "Whether the premium that silver fetched in China over other metals, or for that matter China's trade surplus with Europe before the 1830s, can be viewed as evidence for the 'magnetic' qualities of the Chinese economy should be examined in view of European monetary penetration patterns elsewhere. To understand China's monetary function in the early modern world, we must look beyond silver to the dynamics of monetization following contact with Europe." See Horesh, "Chinese Money in Global Context," 113.

9. Mill, *History of British India*, 1:19. Hosea Morse, who worked for the Imperial Maritime Customs Service in China from 1874 to 1908, estimated that the annual EIC export of silver and coin to "the East Indies" between 1601 and 1620 was £28,847; Morse, *Chronicles of the East India Company*, 8. I leave it to economic historians of Britain to sort out the differences in these estimates.

10. This chapter relies on the version of Mun's text reprinted in Purchas, *Purchas His Pilgrimes*. On Mun's mercantilism, see Kindleberger, *Historical Economics*, 87–100; Harris, *Sick Economies*, 164–68.

11. Mun, *Discourse of Trade*, 268–69, 291–92.

12. Mun, *Discourse of Trade*, 290–91, 293.

13. The exchange rate set in the Hongwu-era price schedule was 1:5. In the Wanli era, Portuguese reports put the value of gold slightly higher, at 5.4 taels of silver, with a price as high as 7 taels for the most refined gold; see Boxer, *Great Ship from Amacon*, 179, 184.

14. Von Glahn, *Fountain of Fortune*, 125–33.

15. Shen Defu in his otherwise thorough account of the Maritime Trade Supervisorate leaves out the Quanzhou office, possibly because it was closed in the Jiajing era; *Wanli yehuo bian*, 317.

16. Guangzhou shi wenwu guanlichu, "Guangzhou dongshan Ming taijian Wei Juan mu qingli jianbao," 282. The rubbing included in the article shows a Venetian ducat, possibly issued under the doge Antonio Vernier late in the fourteenth century.

17. Wang Guangyao, *Mingdai gongting taoci shi*, 224. For the text of this order, see Koo Bumjin, *Imun yŏkchu*, 70–71.

18. I take "paid the prices," *gei jia*, to mean not the price the envoys asked but the price that the Ministry of Rites judged fair; see *Da Ming huidian*, 111.7b–10a.

19. *Ming shizong shilu*, *juan* 68. Note that the term *tiecuo* can also mean "iron files," a confusion that the text does not resolve.

20. Zhang Tingyu, *Ming shi*, 1980.

21. *Da Ming huidian*, 111.7b, 8a, 9b, 10a, 15b, 16a.

22. *Da Ming huidian*, 113.

23. *Ming xiaozong shilu*, 73.3a–b; *Ming shi*, 4867–68; Brook, *Troubled Empire*, 222–23.

24. Memorial of Chen Boxian (27 June 1514), *Ming wuzong shilu* 113.2a, quoted in Brook, *Great State*, 156. For the fuller context, see Brook, "Trade and Conflict in the South China Sea," 26–29.

25. Shen Defu, *Wanli yehuo bian*, 317.

26. Quanzhou shi wenwu guanli weiyuanhui, "Fujian Quanzhou diqu chutu de wupi waiguo yinbi," 373–80. According to Zhang Xie, the largest silver coin circulating in Manila was the *huang bizhi*, literally "gold peso," which had a value of ¾ tael; *Dongxi yang kao*, 94.

27. See Brook, "Merchant Network in 16th Century China," 206–7, in which Zhang is quoted.

28. Lee and Ostigosa, "Studies on the Map *Ku Chin Hsing Sheng Chih Tu*," 6, quoting Gaspar de San Augustin, *Conquistas de las Islas Filipinas*.

29. "Cuentas de las primeras compras que hicieron los oficiales de Manila a los mercaderes Chinos" (Accounts of the first purchases of Manila officials from Chinese merchants), reprinted in Gil, *Los Chinos en Manila*, 561–67. The original document is preserved in Caja de Filipinas: Cuentas de Real Hacienda, Archivo General de Indias, Sevilla (ES.41091.AGI/16/Contaduria 1195). I am grateful to Niping Yan for transcribing and explaining the data. Nothing in the document confirms that the units of taes and maes (taels and mace) corresponded precisely to liang and qian, the units of account in silver used in the Ming. The problem is that the ratio of mae and tae in Manila at the time was 16:1 whereas the ratio of qian to liang in Fujian was 10:1. Soon thereafter Manila reverted to the standard ratio of 10:1.

30. A Spaniard writing from Manila in 1569 notes that rice, pigs, goats, and oxen were available in quantity and at low prices; letter of Martin de Rada to the Marquis de Falces, reprinted in Filipiniana Book Guild, *Colonization and Conquest of the Philippines by Spain*, 149.

31. Writing of the Indian Ocean trading economy, Sebastian Prange has proposed that pepper played a role there similar to the role silver played in the Pacific trading world and that sugar played in the Atlantic; "'Measuring by the Bushel,'" 235. It could be argued that pepper played a similar role in integrating the South China Sea economy, second only to silver.

32. Purchas, *Purchas His Pilgrimes*, 3:504–19; Saris, *Voyage of John Saris*, 202–7. The challenge in using Saris's data lies in sorting out currencies and units of weight and account, which were never the same between one port and the next. For example, textiles are quoted in feet, yards, hastas ("halfe a yard, accounted from your elbow to the toppe of your middle finger"), sasockes ("three quarters of a yard"), bolts ("an hundred and twelve yards the peece"), pieces ("thirteene yards the peece," "twelve yards the peece," "nine yards the peece"), and "Flemmish ells" (for which he gives no precise equivalent, but which I have estimated at 45 in.), as well as by weight in piculs. Spices are measured by catties and bahars. In Ternate, "the Cattee there is three pound five ounces English, the Bahar two hundred Cattees. Item, nineteen Cattees Ternate, makes fifty Cattees Bantam exactly." In Bantam, however, "ten Cattees is an Uta, ten Utaes is a Bahar," whereas in Banda, "a small Bahar is ten Cattees Mace, and a hundred Cattees Nuts; & a great Bahar Mace, is an hundred Cattees, and a thousand Cattees Nuts, and a Cattee is five pound, thirteen ounces and an half English, the prices variable"; see Purchas, *Purchas His Pilgrimes*,

3:511. To establish some consistency, I have recalculated prices on the basis of his report that a Chinese tael was equal to 1⅕ English ounces (36.9 g, close to the tael weight used in this book of 37.3 g), the equivalent of which in Spanish silver coins was 1 7/20 pieces of eight reals. Even that recalculation is not perfect, for Saris notes that the Javanese tael in Bantam was worth 9 to the Chinese 10. Summarizing the relationship between the two tael units, Saris declares that 10 Chinese taels equals 6 Javan taels "exactly"; given the precision of his other data on exchange rates, however, "6" has to be a misprint for "9." These variation in silver and coin weights meant opportunities for arbitrage. Saris advises traders heading for Sukadana to stop first at Banjarmasin, "where you shall have for three Cattees Cashes the Mallaca Taile, which is nine Rialls of eight, as I have been credibly informed, it hath been worth of late years. And bringing it to Soocodanna you shall put it away for Diamonds, at four Cattees Cashes the Taile, which is one and three quarters and half quarter of a Riall in weight, so that you shall gain three quarters of a Riall of eight upon a Taile."

33. That noted, I have retrieved one other price for mercury, in Macau, of 0.53 taels; Boxer, *Great Ship from Amacon*, 180. Competition among European buyers in Macau may have been responsible for this higher price.

34. Velho, *Le premier voyage de Vasco de Gama aux Indes*, 111–16. Prices are recorded in cruzados per quintal. When the cruzado was introduced in the fifteenth century, it had a value of 324 reals. Assigning 1 real a silver weight of 3.3 grams, the silver weight of 1 cruzado is approximately 1.07 kg. According to the record of Ma Huan, an old Portuguese quintal at Calicut weighed roughly 94 catties, or 56 kg, though this figure may be slightly under its standard weight; see Prange, "'Measuring by the Bushel,'" 224n41. The report's attention to prices in Alexandria reflected the Portuguese ambition of displacing the Venetian merchants in Alexandria who dominated the trade in Asian goods; Cook, *Matters of Exchange*, 11.

35. The prices of the other commodities that appear in both the Saris and the Velho lists show a similar drastic collapse: sappanwood in Japan sold for 3 percent of what it had cost in Tenasserim; nutmeg in Japan was down to 1.4 percent from what it had been in Malacca; and benzoin in Bantam cost a mere two-thirds of its old price in Ayutthaya.

36. Torres, "'There Is But One World,'" 2.

37. Boxer, *Great Ship from Amacon*, 179–84. In preference to Boxer's date of 1600, I have dated the list to 1608, the middle date of the reports on East Asian trade Baeza submitted to court between 1607 and 1609; see Torres, "'There Is But One World,'" 13nn2 and 6.

38. Quoted in Gallagher, *China in the Sixteenth Century*, 16–18. I have revised Gallagher's "pounds" and "gold pieces" to catties and silver taels, as gold was not a currency in Ming China.

39. Quoted in Zhang Yi, *Yuguang jianqi ji*, 1010. Sambiasi's Chinese name was Bi Jinji, but he took the studio name of Jinliang, which is how Zhang refers to him.

40. Brook, "Trading Places," 74.

41. Volker, *Porcelain and the Dutch East India Company*, 24–26, 35–45, 227.

42. Lavin, *Mission to China*, 77–79, citing Ruggieri's "Relaciones" in the Archivum Romanum Societatis Iesu (Jap. Sin. 101).

43. In this regard, Chui-mei Ho notes that while only 16 percent of Ming porcelain exports early in the seventeenth century went to Europe, Europe's share was worth 50 percent of the value of that trade; "Ceramic Trade in Asia, 1602–82," 48–49.

44. Shen Defu, *Wanli yehuo bian*, 680. I believe the sentence refers to the prices the envoys paid in Beijing, but it could be construed to mean the prices at which they expected to sell them.

45. "Letter from Fray Martin Ignacio de Loyola," in Blair and Robertson, *Philippine Islands* 12:58–59; some spellings and punctuation have been updated and one minor error corrected. Although Ignacio de Loyola was a grandnephew of the founder of the Society of Jesuits of the same name, he was ordained as a Franciscan and not a Jesuit.

46. Six years later, Pedro de Baeza estimated the amount of silver flowing from Manila to Mexico at 2½ and 3 million silver reals a year, a third the amount Loyola gives; Boxer, *Great Ship from Amacon*, 74.

47. "Letter from Fray Martin Ignacio de Loyola," in Blair and Robertson, *Philippine Islands* 12:60.

48. Zúñiga letters appended to "Letter from Fray Martin Ignacio de Loyola," 12:61–63.

49. Zúñiga letters appended to "Letter from Fray Martin Ignacio de Loyola," 12:64–65.

50. Zúñiga letter, 25 May 1602, appended to "Letter from Fray Martin Ignacio de Loyola," 12:57–75.

51. Fluctuations in the volume of currency affected prices elsewhere in the system, to the distress of merchants holding those currencies. Edmund Scott, an English East India Company agent working in Bantam, observes this effect when a Chinese junk arrived on 22 April 1604, laden with copper cash. When it arrived, the value of copper cash fell in Bantam, which Scott notes because he had set aside a considerable stock of copper coins to make local purchases. The flood of coppers had the effect of both pushing up real prices and driving up the cost of silver reals. See Scott, *Exact Discourse*, E1.

52. Boyd-Bowman, "Two Country Stores in XVIIth Century Mexico," 242, 244, 247.

53. Yan Junyan, *Mengshui zhai cundu*, 702.

54. Gu Yanwu, *Tianxia junguo libing shu*, 26.33a–34a. Fu's intervention is examined in Pin-tsun Chang, "Sea as Arable Fields," 20, 24–25.

55. *Jinjiang xianzhi* (1765), 10.70b–71a; see also 8.58b for Fu Yuanchu's name in the list of successful examination graduates. Xue Longchun offers a few details about Fu in *Wang Duo nianpu changbian*, entries for seventeenth day, eleventh month, 1637, and fourteenth day, first month, 1638. Fu's dismissal is noted in Zhang Tingyu, *Ming shi*, 6863; see also 6672.

56. Gu Yanwu, *Tianxia junguo libing shu*, 26.34a.

57. The debate resurfaced in a more vociferous second round in the 1660s; Chaudhuri, *Trading World of Asia and the East India Company*, 8.

58. The clearest correction to this argument is von Glahn, *Fountain of Fortune*, 113–41.

59. Crosby, *Columbian Exchange*.

60. Hamilton, "American Treasure and the Rise of Capitalism," 349–57.

61. Cited in Munro, "Money, Prices, Wages, and 'Profit Inflation,'" 15.

62. Munro, "Money, Prices, Wages, and 'Profit Inflation,'" 18.

Chapter 4. The Famine Price of Grain

1. As recorded by the official diarist of the embassy, Ghiyasu'd-Din Naqqash, in Abru, *Persian Embassy to China*, 62.

2. On Yongle's attempts to ease his usurper's status, see Brook, *Great State*, 85–88.

3. The only price records I have recovered for the Yongle Slough are in 1404 and 1405: *Raozhou fuzhi* (1872), 31.29b; *Qian shu* (1618), repeated in *Qianshan xianzhi* (1784), 1.

4. E.g., Wu Hong, *Zhishang jinglun*, 6.3b.

5. *Da Ming huidian*, 179.4a, converting prices in scrip to silver and copper at their official rates of exchange.

6. Boxer, *Great Ship from Amacon*, 184–85.

7. Wan Shihe, "Tiaochen nanliang quefa shiyi shu" (Memorial on measures to deal with shortages in grain tribute from the south), in *Wan Wengong gong zhaiji*, 11.8b, quoted in Kishimoto, *Shindai Chūgoku no bukka*, 226.

8. Zhao Yongxian, "Qing ping Jiangnan liangyi shu" (Memorial requesting a leveling of taxes in Jiangnan), *Huang Ming jingshi wenbian*, 397.9a–b. Zhao makes this observation in the context of working out favorable rates for converting grain taxes into silver payments.

9. Zhao Yongxian, *Songshi zhai ji*, 27.8a, 10b.

10. Ren Yuanxiang, quoted in Kishimoto, *Shindai Chūgoku no bukka*, 226.

11. Tang Shunzhi, "Yu Li Longgang yiling" (Letter to county magistrate Li Longgang), *Tang Jingchuan xiansheng wenji*, 9.

12. Lu Wenheng, *Se'an suibi*, 3.5a, quoted in Kishimoto, *Shindai Chūgoku no bukka*, 230.

13. Liu Benpei, quoted in Kishimoto, *Shindai Chūgoku no bukka*, 230.

14. *Jiading xianzhi* (1881), 3.13b.

15. Other grains continued to be cheaper than rice later in the dynasty, as they had been in 1368, though in this chapter the focus of the narrative is on the price of rice.

16. Qing historiographer Zhang Xuecheng regarded the lack of price data in gazetteers as a failing and advised compilers to include grain and commodity prices in their books; Wilkinson, *Studies in Chinese Price History*, 2, 5. Even rarer than prices in gazetteers are reports of weather; a notable exception is *Shaowu fuzhi* (1543), 1.5b–11a.

17. E.g., in chronological order between 1568 and 1589: *Guihua xianzhi* (1614), 10; *Haicheng xianzhi* (1633), 14.2a; *Fujian tongzhi* (1871), 271.34a; *Hangzhou fuzhi* (1922), 84.23a; *Luzhou fuzhi* (1885), 93.10b.

18. E.g., *Leting xianzhi* (1755), 12.13a; *Hangzhou fuzhi* (1784), 56.17a; *Yan'an fuzhi* (1802), 6.1b; *Yongping fuzhi* (1879), 30.26a.

19. Examples of 30 coppers between 1613 and 1618: *Xiongxian xinzhi* (1930), 8.45b; *Qihe xianzhi* (1673), 6; *Jinan fuzhi* (1840), 20.17b.

20. Zhang Tingyu, *Ming shi*, 1.

21. Zhu Yuanzhang, *Ming taizu ji*, 153. Emperor Yongle cited the same Tang price in response to a memorial in 1422 reporting on a famine north of the Yangzi River; *Ming taizong shilu*, 247.1b.

22. Brook, *Confusions of Pleasure*, 70–71.

23. *Guangzhou zhi* (1660), 11.26a.

24. Xie Qian, "Lianghuai shuizai qi zhenji shu" (Memorial seeking relief for the food disaster in the Huai region), in Chen Zilong, *Huang Ming jingshi wenbian*, 97.9b.

25. *Pingyuan xianzhi* (1749), 9.7b.

26. *Songjiang fuzhi* (1630), 13.74a. The price of rice rose to 130 coppers per peck that year, well above the normal famine price of 100 coppers, but below the prices that were starting to move into the hundreds of coppers in the decade. The prefect is not mentioned in the brief gazetteer record, but his leadership can be assumed. For a case in 1638 when the local elite failed to make their stocks of grain available, provoking an uprising, see *Wujiang xianzhi* (1747), 40.32b.

27. On the relationship between state intervention and commercial supply in the Ming, see Brook, *Confusions of Pleasure*, 102–4, 190–93.

28. Wu Yingji, *Liudu jianwen lu*, quoted in Qin Peiheng, "Mingdai mijia kao," 204.

29. For reports of famine cannibalism in 1523 and 1589, see *Yancheng xianzhi* (1875), 17.2b; *Luzhou fuzhi* (1885), 93.10b. For reports of cannibalism from the 1615–16 famine in Shandong, see Xu Hong, "Jieshao jize Wanli sishisan."

30. *Yuanwu xianzhi* (1747), 10.5a, 6a.

31. *Wenxiang xianzhi* (1932), 1.5b, 6a.

32. *Xiajin xianzhi* (1741), 9.9b.

33. *Gushi xianzhi* (1659), 9.24b; on the place of millet in county agriculture, see 2.25a.

34. As an editor of the Zhejiang provincial gazetteer explained to readers, deciding whether to record only auspicious signs or only omens (the latter practice being a tradition going back to the *Spring and Autumn Annals*), or both, put compilers in the difficult position of having to manage the political impact of recording the truth. He concludes that the best course is just to record everything and let readers deal with it; *Zhejiang tongzhi* (1561), 63.17b–18a.

35. E.g., *Pinghu xianzhi* (1627), 18.23a.

36. Fujian official Lin Xiyuan, influential in famine policy in the mid-sixteenth century, allowed merchants to add 2 cents per hectoliter, half to cover transport costs and half to pay themselves a commission. Buyers absorbed the surcharge and the total cost to the government of relieving a famine was nothing; cited in Lu Zengyu, *Kangji lu*, 3a.48a–b. The author credits Lin Xiyuan and Wang Shangjiong as the two principal architects of Ming famine relief policies.

37. *Haicheng xianzhi* (1633), 18.5b.

38. A team of disaster historians under the direction of Jia Guirong and Pian Yuqian is photoreproducing the disasters sections of gazetteers in hope of producing a complete archive of China's historical disasters, under the general title of *Difangzhi zaiyi ziliao congkan* (Compendium of materials on disasters from local gazetteers).

39. Brook, "Spread of Rice Cultivation," 660, following the analysis of John Lossing Buck. Buck constructed his division between the cultivation of gaoliang and winter wheat in north China and rice in the center and south. In the Ming, the principal food grain in north China was millet, not gaoliang, but the geographical boundary that Buck identified is still applicable.

40. For this exercise, I have used the provincial totals, adjusted for Ming boundaries, in Wang Teh-yi, *Zhonghua minguo Taiwan diqu gongcang fangzhi mulu*, 1–98, 101–239.

41. For provincial populations, see Li Defu, *Mingdai renkou yu jingji fazhan*, 127, taken from the official (and not hugely reliable) figures given in the 1578 census, for which see Liang Fangzhong, *Zhongguo lidai hukou, tiandi, tianfu tongji*, 341.

42. C. Dyer, *Standards of Living in the Later Middle Ages*, 264. On the relationship between climate and grain prices in south China, see Marks, "'It Never Used to Snow.'"

43. The first comprehensive study of climate disasters using the dynastic histories and the Qing encyclopedia *Tushu jicheng* (Compendium of texts and images) is Chen Gaoyong, *Zhongguo lidai tianzai renhuo biao*. Among later compendia, Satō Taketoshi adds data from the imperial biographies in the dynastic histories in his *Chūgoku saigaishi nenpyō*. Song Zhenghai also uses the dynastic histories, plus data from the *Veritable Records* and local gazetteers, in his *Zhongguo gudai ziran zaiyi xiangguanxing nianbiao zonghui*.

44. Song Lian, *Yuan shi*, 1053.

45. The seriousness with which court historians took the charge of producing history-worthy records lends consistency to their reports of climate disturbances. Some readers complained, as Long Wenbin does in the nineteenth century, that "the two chapters of the Five Phases section of the *Ming shi* could not entirely record all categories of disaster (*Ming huiyao*, 4), yet he found that he was unable to supplement the official record.

46. At the provincial level, *Anhui tongzhi* (1877), *juan* 347; *Fujian tongzhi* (1871), *juan* 271; *Gansu xin tongzhi* (1909), *juan* 2; *Hubei tongzhi* (1921), *juan* 75; *Hunan tongzhi* (1885), *juan* 243; *Shanxi tongzhi* (1734), *juan* 30; *Sichuan tongzhi* (1816), *juan* 203; and *Zhejiang tongzhi* (1735), *juan* 109. At the prefectural level, *Jinan fuzhi* (1840), *juan* 20; *Linqing zhouzhi* (1674), *juan* 3; *Songjiang fuzhi* (1630), *juan* 47; *Suzhou fuzhi* (1642); *Yunzhong zhi* (1652), *juan* 12; and *Zhending fuzhi* (1762), *juan* 7. The south and southwest are underrepresented in this sample.

47. Grove, "Onset of the Little Ice Age," 160–62.

48. See Siebert et al., *Volcanoes of the World*, 239, 324. See also Atwell, "Volcanism and Short-Term Climate Change," 50–55.

49. *Jiangdu xianzhi* (1881), 2.13b. In "Time, Money, and the Weather," 84–85, Atwell offers a slightly warmer climate profile for the fifteenth century by drawing on summer anomalies taken from Northern Hemisphere locations outside China.

50. Gallagher, *China in the Sixteenth Century*, 14, 316.

51. On the 1590s as a cold decade throughout the Northern Hemisphere, see Parker, "History and Climate."

52. This late cold phase has been widely noted; e.g., Zhang Jiacheng and Crowley, "Historical Climate Records in China and Reconstruction of Past Climates," 841. In *The Reconstruction of Climate in China for Historical Times*, 98, 107, Zhang Jiacheng proposes a larger pattern of cold and warm "ages" that does not tally with my findings; nor do my data support his hypothesis of a regular, periodic pattern of drought phases. According to Ge Quansheng, yearly temperatures during this period were depressed by roughly 1° C, and summer temperatures by 2° C; Ge Quansheng et al., "Coherence of Climatic Reconstruction," 1014.

53. This finding is confirmed by the annual precipitation maps compiled by the Central Hydrological Bureau, based on reports taken from local gazetteers since 1470; Zhongyang qixiangju qixiang kexue yanjiu yuan, *Zhongguo jin wubai nian hanlao fenbu tuji*.

54. This profile correlates closely with Europe's; see Alexandre, *Le climat en Europe*, 776–82.

55. *Shaoxing fuzhi* (1586), 13.32b.

56. Zhang Tingyu, *Ming shi*, 485.

57. On the correlation of precipitation swings with the El Niño Southern Oscillation (ENSO), see Brook, "Nine Sloughs," 43–45.

58. This famine is surveyed in Dunstan, "Late Ming Epidemics," 8–18.

59. *Ming shenzong shilu*, 197.3a, 11a.

60. Le Roy Ladurie, *Histoire humaine et comparée du climat*, 225–37.

61. Marks, *China: Its Environment and History*, 188, tags 1614 as a turning point in south China.

62. *Ming shenzong shilu*, 538.2b, 539.9b, 540.7b, 542.2b.

63. Gu Qiyuan, *Kezuo zhuiyu*, 1.30b; Wu Yingji, *Liudu jianwen lu*, 2.13b.

64. Bauernfeind and Woitek, "Influence of Climatic Change," 307, 320.

65. On the difficulty of aligning the local instantiation of large-scale trends with climate, see Brook, "Differential Effects of Global and Local Climate Data."

66. *Songjiang fuzhi* (1818), 80.18b.

67. The Chongzhen drought is explored in Zheng et al., "How Climate Change Impacted the Collapse of the Ming Dynasty." In support of the argument that climate anomalies precipitated the fall of the Ming, see H. Cheng et al., "Comment on 'On Linking Climate to Chinese Dynastic Change.'"

68. *Songjiang fuzhi* (1818), 80.18b–20a.

69. For a broad-brush attempt to model the impact of climate on human disasters in this period, see Xiao et al., "Famine, Migration and War."

70. *Neiqiu xianzhi* (1832), 3.44b–45b.

71. Parker, *Global Crisis*, 3–8 passim.

72. Siebert et al., *Volcanoes of the World*, 244–45, 324. See also Atwell, "Volcanism and Short-Term Climate Change," 62–70.

Chapter 5. The Chongzhen Price Surge

1. Munro, "Money, Prices, Wages, and 'Profit Inflation,'" 17–18.

2. Here I loosely follow Fischer's summary in the conclusion of *The Great Wave*, 237–39.

3. Munro, review of Fischer, *The Great Wave*.

4. Ye Mengzhu, *Yueshi bian*, 15.

5. This assumption is complicated by the revision around 1373 of the value of one string down to 400 coppers for legal purposes; Farmer, *Zhu Yuanzhang and Early Ming Legislation*, 186. However, I see no evidence that this revision has been applied to the 1451 list.

6. There are several modern reprints of *Tianshui bingshan lu*, but no critical edition. I have relied on the 1951 edition in *Ming wuzong waiji* (ed. Mao Qiling, reprinted in 1982); the prices are on pages 157–64, 170. On the reputation of the text, see Dardess, *Four Seasons*, 215; Clunas, *Superfluous Things*, 46–49; also Wu Renshu, *Pinwei shehua*, 233–37.

7. Confirmed by Kishimoto, *Shindai Chūgoku no bukka*, 220–27. Li Defu makes a similar argument in the context of population growing against a background of mild deflation in the sixteenth century; *Mingdai renkou yu jingji fazhan*, 89–101.

8. Hai Rui, *Hai Rui ji*, 42.

9. *Chengshi randian chasuan zhangpu*, reproduced in *Huizhou qiannian qiyue wenshu*.

10. Li Guimin, "Ming-Qing shiqi landianye yanjiu," 150. Li's discussion of the ledger at 144–50 is focused on capital formation rather than prices.

11. Ye Mengzhu, *Yueshi bian*, 159. The book remained in manuscript until a Shanghai publisher got hold of a copy in the Songjiang Library and published it in 1934. Since the 1970s, it has been a favorite of price historians.

12. Ye Mengzhu, *Yueshi bian*, 153.

13. The Shanghai county gazetteer blames the collapse of copper in 1642 on private debasement; *Shanghai xianzhi* (1882), 30.9b. In an earlier collapse in the copper-silver exchange rate in Suzhou in 1599, it took 3,000 coppers to reach 1 tael; *Wuxian zhi* (1642), 2.30b–31a.

14. Atwell, "International Bullion Flows and the Chinese Economy," 88. For a well-grounded challenge to this interpretation, see von Glahn, "Changing Significance of Latin American Silver in the Chinese Economy," 558–61.

15. Ye Mengzhu, *Yueshi bian*, 14–15.

16. Ye Mengzhu, *Yueshi bian*, 15.

17. Ye Mengzhu, *Yueshi bian*, 153–61.

18. Ye Mengzhu, *Yueshi bian*, 153–54.

19. Von Glahn, *Fountain of Fortune*, 215–16.

20. The use of the later benchmark to construct a price curve in the seventeenth century dominates the analysis in, among other studies, Quan Hansheng, "Song Ming jian baiyin goumaili de biandong," 165; Y. Wang, "Secular Trends of Rice Prices in the Yangtze Delta," 39–40; Kishimoto Mio, *Shindai Chūgoku no bukka*, 112–16; von Glahn, "Money Use in China and Changing Patterns of Global Trade in Monetary Metals," 191–92.

21. On the collapse of textile prices in Songjiang, see Ye Mengzhu, *Yueshi bian*, 157–58.

22. Parker, *Global Crisis*, ch. 5.

23. Le Roy Ladurie, *Histoire humaine et comparée du climat*, 100.

24. E.g., Liu Jian et al., "Simulated and Reconstructed Winter Temperatures," 2875.

25. *Yangzhou fuzhi* (1733).

26. Y. Wang, "Secular Trend of Prices during the Ch'ing Period," 362. The crisis is examined on a global scale in Davis, *Late Victorian Holocaust*.

Afterword. Climate and History

1. Quoted in Brook, *Troubled Empire*, 240.

2. For example, changes in irrigation technology through the twentieth century profoundly altered agricultural stability; Kueh, *Agricultural Instability in China*, 257–58.

3. Brook, *Confusions of Pleasure*, 103–4.

4. Li Zichun, "Ming mo yijian youguan wujia de shiliao."

BIBLIOGRAPHY

Note: Local gazetteers are cited in the notes by title and date of edition but do not appear in this bibliography.

Primary Sources

Abru, Hafiz, ed. *A Persian Embassy to China*. Translated by K. M. Maitra. Edited by L. Carrington Goodrich. New York: Paragon, 1970.

Blair, Helen, and James Robertson, eds. *The Philippine Islands, 1493–1803*. 55 vols. Cleveland: Arthur H. Clark, 1903–9.

Chen Qide 陳其德. *Chuixun puyu* 垂訓朴語 [Simple words handed down to instruct]. Edited by Chen Zi 陳梓. Tongxiang, 1813.

Chen Zilong 陳子龍, ed. *Huang Ming jingshi wenbian* 皇明經世文編 [Collected essays on statecraft from the imperial Ming]. 1638. Beijing: Zhonghua shuju, 1987.

Chengshi randian chasuan zhangbu 程氏染店查算帳簿 [Annual ledger of Cheng's Dyeworks]. 1594–1604. Reprinted in *Huizhou qiannian qiyue wenshu* 徽州千年契約文書 [A thousand years of contract documents from Huizhou], edited by Wang Yuxin 王鈺欣 and Zhou Shaoquan 周绍泉, 8:74–158. Shijiazhuang: Huashan wenyi chubanshe, 1991–93.

Ch'oe Pu. *A Record of Drifting across the Sea*. Translated by John Meskill. Tucson: University of Arizona Press, 1965.

Chuan zheng [Barge administration]. Nanjing: Ministry of War, 1546.

Da Ming huidian 大明會典 [Collected statutes of the Ming Great State]. Beijing, 1587.

Da Yuan shengzheng guochao dianzhang 大元聖政國朝典章 [Dynastic statutes of the sagely administration of the Yuan Great State]. 1322. Taipei: Guoli gugong bowuyuan, 1976.

Dean, Kenneth, and Zheng Zhenman, eds. *Fujian zongjiao beiwen huibian: Quanzhou fu fence* [Collected epigraphy from Fujian religious institutions: Quanzhou prefecture], 3 parts. Fuzhou: Fuzhou renmin chubanshe, 2003.

———. *Fujian zongjiao beiwen huibian: Xinghua fu fence* [Collected epigraphy from Fujian religious institutions: Xinghua prefecture]. Fuzhou: Fuzhou renmin chubanshe, 1995.

Fan Lai 范來. *Liangzhe haifang leikao xubian* 兩浙海防類考續編 [Further compendium on the maritime defense of Zhejiang]. 1602.

Feng Menglong 馮夢龍. *Stories Old and New: A Ming Dynasty Collection*. Translated by Shuhui Yang and Yinqin Yang. Seattle: University of Washington Press, 2000.

Feng Mengzhen 馮夢禎. *Kuaixue tang riji* 快雪堂日記 [Notes from the Hall for Taking Pleasure in the Snow]. Nanjing: Fenghuang chubanshe, 2010.

Ge Yinliang 葛寅亮, ed. *Jinling fancha zhi* 金陵梵刹志 [Gazetteer of the Buddhist monasteries of Nanjing]. Nanjing: Ministry of Rites, 1607.

Girard, Pascal, trans. *Le voyage en Chine d'Adriano de las Cortes S.J. (1625)*. Paris: Chandeigne, 2001.

Gu Qiyuan 顧起元. *Kezuo zhuiyu* 客座贅語 [Trivial comments from the guest's seat]. Nanjing, 1618.

Gu Yanwu 顧炎武. *Tianxia junguo libing shu* 天下郡國利病書 [Strengths and weaknesses of the regions of the realm]. Kyoto: Chūmon shuppansha, 1975.

Hai Rui 海瑞. *Hai Rui ji* 海瑞集 [Collected writings of Hai Rui]. Beijing: Zhonghua shuju, 1981.

Jia Guirong 賈貴榮 and Pian Yuqian 駢宇騫, eds. *Difangzhi zaiyi ziliao congkan* 地方志災異資料叢刊 [Collection of materials on natural disasters in local gazetteers]. Beijing: Guojia tushuguan chubanshe, 2010.

Koo Bumjin [Ku Pŏm-jin], ed. *Imun yŏkchu* 吏文譯註 [Annotated translation of "Clerical texts"]. Seoul: Ch'ulp'ansa, 2012.

Li Le 李樂. *Jianwen zaji* 見聞雜記 [Notes on things I have seen and heard]. 1610, with supplements to 1612. Shanghai: Shanghai guji chubanshe, 1986.

Li Rihua 李日華. *Weishui xuan riji* 味水軒日記 [Diary from Water-Tasting Pavilion]. Shanghai: Yuandong chubanshe, 1996.

Liu Ruoyu 劉若愚. *Zhuozhong zhi* 酌中志 [A weighted and unbiased record]. Beijing: Beijing chubanshe, 2000.

Long Wenbin 龍文彬. *Ming huiyao* 明會要 [Digest of Ming statutes]. 1887. Beijing: Zhonghua shuju, 1956.

Lü Kun 呂坤. *Shizheng lu* 時政錄 [A record of contemporary administration]. 1598. Taipei: Wenshizhe chubanshe, 1971.

Lu Wenheng 陸文衡. *Se'an suibi* 嗇菴隨筆 [Jottings from Se Hermitage]. Taipei: Guangwen shuju, 1969.

Lu Zengyu 陸曾禹. *Kangji lu* 康濟錄 [A record of aiding the living]. 1739. Suzhou, 1784.

Mao Qiling 毛奇齡, ed. *Ming wuzong waiji* 明武宗外記 [Unofficial records of the Zhengde era]. 1947. Beijing: Shenzhou guoguang she, 1951; Shanghai: Shanghai shudian, 1982.

Mill, James. *The History of British India*. 3 vols. London: Baldwin, Cradock, and Joy, 1817.

Ming chongzhen changbian 明崇禎長編 [Unedited records of the Chongzhen reign]. Taipei: Taiwan yinhang, 1969.

Ming shenzong shilu 明神宗實錄 [Veritable records of the Wanli reign]. Taipei: Zhongyang yanjiuyuan lishi yuyan yanjiusuo, 1962.

Ming shizong shilu 明世宗實錄 [Veritable records of the Jiajing reign]. Taipei: Zhongyang yanjiuyuan lishi yuyan yanjiusuo, 1962.

Ming taizong shilu 明太宗實錄 [Veritable records of the Yongle reign]. Taipei: Zhongyang yanjiuyuan lishi yuyan yanjiusuo, 1962.

Ming taizu shilu 明太祖實錄 [Veritable records of the Hongwu reign]. Taipei: Zhongyang yanjiuyuan lishi yuyan yanjiusuo, 1962.

Ming wuzong shilu 明武宗實錄 [Veritable records of the Zhengde reign]. Taipei: Zhongyang yanjiuyuan lishi yuyan yanjiusuo, 1962.

Ming xiaozong shilu 明孝宗實錄 [Veritable records of the Hongzhi reign]. Taipei: Zhongyang yanjiuyuan lishi yuyan yanjiusuo, 1962.

Ming xuanzong shilu 明宣宗實錄 [Veritable records of the Xuande reign]. Taipei: Zhongyang yanjiuyuan lishi yuyan yanjiusuo, 1962.

Mun, Thomas. *A Discourse of Trade, from England unto the East-Indies: Answering to Diverse Objections Which Are Usually Made against the Same.* 1621. 2nd ed. (1621), reprinted in *Purchas His Pilgrimes*, edited by Samuel Purchas, 5:262–301. Reprinted in *A Select Collection of Early English Tracts on Commerce*, edited by J. R. McCulloch, 1–47. London: Political Economy Club, 1856.

Pantoja, Diego. *Advis du Reverend Père Iaques Pantoie de la Compagnie de Jésus envoyé de Paquin Cité de la Chine.* Translation of *Relacion de la Entrada de Algunos Padres de la Compañia de Iesus en la China.* Arras: Guillaume de la Rivière, 1607.

Purchas, Samuel. *Purchas His Pilgrimes: Contayning a History of the World in Sea Voyages and Lande Travells by Englishmen and Others.* London: Henrie Featherstone, 1625. Glasgow: James MacLehose and Sons, 1905.

Saris, John. *The Voyage of Captain John Saris to Japan, 1613.* Edited by Ernest Satow. London: Hakluyt Society, 1900.

Scott, Edmund. *An Exact Discourse of the Subtleties, Fashions, Policies, Religion, and Ceremonies of the East Indians.* London: Walter Burre, 1606.

Shanghai bowuguan tushu ziliao shi 上海博物館圖書資料室 [Publications office of the Shanghai Museum]. *Shanghai beike ziliao xuanji* 上海碑刻資料選輯 [Selected epigraphic materials from Shanghai]. Shanghai: Renmin chubanshe, 1980.

Shen Bang 沈榜. *Wanshu zaji* 宛署雜記 [Unsystematic records from the Wanping county office]. 1593. Beijing: Beijing guji chubanshe, 1980.

Shen Defu 沈德符. *Wanli yehuo bian* 萬曆野獲編 [Private gleanings from the Wanli era]. 1606. Beijing: Zhonghua shuju, 1997.

Shen shi 沈氏 (Master Shen). *Bu nongshu* 補農書 [Supplement to the Treatise on Agriculture]. Chongzhen era. Reprinted in Zhang Lixiang 張履祥, *Yangyuan xiansheng quanji* 楊園先生全集 [Complete works of Master Yangyuan], 1782.

Shum Chun (Shen Jin 沈津). "Mingdai fangke tushu zhi liutong yu jiage" 明代坊刻圖書之流通與價格 [Circulation and prices of commercially printed books in the Ming period]. *Guojia tushuguan guankan* 國家圖書館館刊 1 (1996): 101–18.

Song Lian 宋濂, ed. *Yuan shi* 元史 [History of the Yuan dynasty]. Beijing: Zhonghua shuju, 1976.

Song Yingxing 宋應星. *Tiangong kaiwu* 天工開物 [The making of things by Heaven and humankind]. Edited by Dong Wen 董文. 1962. Taipei: Shijie shuju, 2002.

Sun, E-tu Zen, and Shiou-chuan Sun, trans. *Chinese Technology in the Seventeenth Century: T'ien-kung k'ai-wu.* University Park: Pennsylvania State University Press, 1966.

Tan Qian 談遷. *Zaolin zazu* 棗林雜俎 [Various offerings from Date Grove]. Beijing: Zhonghua shuju, 2007.

Tang Shunzhi 唐順之. *Tang Jingchuan xiansheng wenji* 唐荊川先生文集 [Collected writings of Master Tang Jingchuan]. 1573.

Tang Zhen 唐甄. *Qian shu* 潛書 [Writings while out of sight]. Beijing: Zhonghua shuju, 1963.

Tianshui bingshan lu 天水冰山錄 [Heaven turning a glacier to water]. 1562. Reprinted in Mao Qiling, *Ming wuzong waiji* [Unofficial records of the Zhengde era], 1951.

Tu Long 屠隆. "Huangzheng kao" 荒政考 [A study of famine administration]. In *Zhongguo huangzheng quanshu* 中國荒政全書, vol. 1, edited by Li Wenhai 李文海 and Xia Mingfang 夏明方, 175–95. Beijing: Beijing guji chubanshe, 2003.

Velho, Álvaro (attrib.). *Le premier voyage de Vasco de Gama aux Indes (1497–1499)*. Paris: Chandeigne, 1998.

Wan Shihe 萬士和. *Wan Wengong gong zhaiji* 萬文恭公摘集 [Selections from the writings of Master Wan Wengong]. 1592.

Wang Shiqiao 王士翹. *Xiguan zhi* 西關志 [Gazetteer of the Western Passes]. 1548. Beijing: Beijing guji chubanshe, 1990.

Wang Yangming. *Instructions for Practical Living, and Other Neo-Confucian Writing*. Translated by Wing-tsit Chan. New York: Columbia University Press, 1963.

Wu Gang 吳鋼, ed. *Huashan beishi* 華山碑石 [Epigraphic records of Mount Hua]. Xi'an: Sanqin chubanshe, 1995.

Wu Hong 吳宏. *Zhishang jinglun* 紙上經綸 [Grand schemes on paper]. Huizhou, 1721. Ōki collection, Tōyō bunka kenkyūjo, University of Tokyo.

Wu Yingji 吳應箕. *Liudu jianwen lu* 留都見聞錄 [Things seen and heard while sojourning in the [southern] capital]. 1644, 1730. Guichi xianzhe yishu reprint, 1920.

Xie Bin 謝彬. *Nanjing hubu zhi* 南京戶部志 [Gazetteer of the Nanjing ministry of revenue]. 1550.

Xie Zhaozhe 謝肇淛. *Wu zazu* 五雜俎 [Fivefold miscellany]. Shanghai: Shanghai shudian chubanshe, 2001.

Xu Guangqi 徐光啟. *Nongzheng quanshu jiaozhu* 農政全書校注 [Complete handbook on agricultural administration, annotated]. 3 vols. Edited by Shi Shenghan 石聲漢. Shanghai: Shanghai guji chubanshe, 1979.

———. *Xu Guangqi ji* 徐光啟集 [Collecting writings of Xu Guangqi]. Shanghai: Shanghai guji chubanshe, 1963.

Yan Junyan 顏俊彥. *Mengshui zhai cundu* 盟水齋存讀 [Notes from Mengshui Studio]. 1632.

Yang Dongming 楊東明. *Jimin tushuo* 飢民圖說 [Album of the famished]. 1748.

Ye Chunji 葉春及. *Huian zhengshu* 惠安政書 [Administrative handbook of Huian]. 1573. Shidong wenji edition, 1672.

Ye Mengzhu 葉夢珠. *Yueshi bian* 閱世編 [A survey of the age]. Shanghai: Shanghai guji chubanshe, 1981.

Yu Sen 俞森. *Huangzheng congshu* 荒政叢書 [Collected writings on famine administration]. 1690.

Yuan Zhongdao 袁中道. *Youju feilu* 遊居柿錄 [Notes made while traveling and in repose]. Shanghai: Yuandong chubanshe, 1996.

Zhang Dai 張岱. *Langhuan wenji* 琅嬛文集 [Collected essays from Langhuan]. Shanghai: Guangyi shuju, 1936.

———. *Taoan mengyi* 桃庵夢遺 [Dream recollections from Tao Hermitage]. Shanghai: Shangwu yinshuguan, 1939.

Zhang Han 張瀚. *Songchuang mengyu* 松窗夢語 [Dream recollections at the pine window]. 1593. Reprinted with *Zhishi yuwen*. Beijing: Zhonghua shuju, 1985.

Zhang Kentang 張肯堂. *Xunci* 礐辭 [Level-field judgments]. 1634. Taipei: Xuesheng shuju, 1970.

Zhang Lixiang 張履祥. *Bu nongshu jiaoshi* 補農書校釋 [Supplement to the *Manual of agriculture*, annotated]. Edited by Chen Hengli 陳恆利. Beijing: Nongye chubanshe, 1983.

———. *Yangyuan xiansheng quanji* 楊園先生全集 [Complete works of Master Yangyuan]. 1782.

Zhang Tingyu 張廷玉, ed. *Ming shi* 明史 [History of the Ming]. Beijing: Zhonghua shuju, 1974.

Zhang Xie 張燮. *Dongxi yang kao* 東西洋考 [Studies of the eastern and western sea routes]. Beijing: Zhonghua shuju, 1981.

Zhang Yi 張怡. *Yuguang jianqi ji* 玉光劍氣集 [The jade-bright sword collection]. Beijing: Zhonghua shuju, 2006.

Zhao Yongxian 趙用賢. *Songshi zhai ji* 松石齋集 [Collection from Songshi Studio]. 1618.

Zhu Fengji 朱逢吉. *Mumin xinjian* 牧民心鑑 [Mirror of the mind for shepherding the people]. 1404. Japanese ed., 1852.

Zhu Yuanzhang 朱元璋. *Ming taizu ji* 明太祖集 [Writings of the founding Ming emperor]. Edited by Hu Shi'e 胡士萼. Hefei: Huangshan shuse, 1991.

Secondary Sources

Agøy, Erling. "Weather Prognostication in Late Imperial China as Presented in Local Gazetteers (1644–1722)." Unpublished.

Alexandre, Pierre. *Le climat en Europe au Moyen Age: Contribution à l'histoire des variations climatiques de 1000 à 1425, d'après les sources narratives de l'Europe occidentale*. Paris: École des Hautes Études en Sciences Sociales, 1987.

Allen, Robert. *The British Industrial Revolution in Global Perspective*. New York: Cambridge University Press, 2009.

Atwell, William. "Another Look at Silver Imports into China, ca. 1635–1644." *Journal of World History* 16, no. 4 (2005): 467–89.

———. "International Bullion Flows and the Chinese Economy circa 1530–1630." *Past and Present* 95 (1982): 68–90.

———. "Notes on Silver, Foreign Trade, and the Late Ming Economy." *Ch'ing-shih wen-t'i* 3, no. 8 (1977): 1–33.

———. "Time, Money, and the Weather: Ming China and the 'Great Depression' of the Mid-Fifteenth Century." *Journal of Asian Studies* 61, no. 1 (February 2002): 83–113.

———. "Volcanism and Short-Term Climate Change in East Asian and World History, c. 1200–1699." *Journal of World History* 12, no. 1 (2001): 29–98.

Bauernfeind, Walter, and Ulrich Woitek. "The Influence of Climatic Change on Price Fluctuations in Germany during the 16th Century Price Revolution." *Climatic Change* 43, no. 1 (1999): 303–21.

Beveridge, William. *Wages and Prices in England from the Twelfth to the Nineteenth Century*. 1939. London: Cass, 1965.

Bian, He. *Know Your Remedies: Pharmacy and Culture in Early Modern China*. Princeton, NJ: Princeton University Press, 2020.

Bol, Peter. *Localizing Learning: The Literati Enterprise in Wuzhou, 1100–1600*. Cambridge, MA: Harvard University Asia Center, 2022.

Boxer, C. R. *South China in the Sixteenth Century*. London: Hakluyt Society, 1953.

———. *The Great Ship from Amacon*. Lisbon: Centro de Estudoes Históricos Ultramarinos, 1960.

Boyd-Bowman, Peter. "Two Country Stores in XVIIth Century Mexico." *Americas* 28, no. 3 (January 1972): 237–51.

Braudel, Fernand. *The Structures of Everyday Life: The Limits of the Possible*. Translated by Siân Reynolds. Civilization and Capitalism, 15th–18th Century 1. London: Collins, 1981.

Britnell, Richard. "Price-Setting in English Borough Markets." *Canadian Journal of History* 31, no. 1 (April 1996): 1–15.

Brook, Timothy. *The Confusions of Pleasure: Commerce and Culture in Ming China*. Berkeley: University of California Press, 1998.

———. "Differential Effects of Global and Local Climate Data in Assessing Environmental Drivers of Epidemic Outbreaks." *Proceedings of the National Academy of Sciences* 114, no. 49 (5 December 2017): 12845–47.

———. *Great State: China and the World*. New York: HarperCollins, 2020.

———. "Great States." *Journal of Asian Studies* 75, no. 4 (November 2016): 957–72.

———. "The Merchant Network in 16th Century China: A Discussion and Translation of Zhang Han's 'On Merchants.'" *Journal of the Economic and Social History of the Orient* 24, no. 2 (1981): 165–214.

———. "Native Identity under Alien Rule: Local Gazetteers of the Yuan Dynasty." In *Pragmatic Literacy, East and West, 1200–1330*, edited by Richard Britnell, 235–45. Woodbridge: Boydell and Brewer, 1997.

———. "Nine Sloughs: Profiling the Climate History of the Yuan and Ming Dynasties, 1260–1644." *Journal of Chinese History* 1 (2017): 27–58.

———. "Something New." In *Early Modern Things: Objects and Their Histories, 1500–1800*, edited by Paula Findlen, 369–74. Abingdon: Routledge, 2013.

———. "The Spread of Rice Cultivation and Rice Technology into the Hebei Region in the Ming and Qing." In *Explorations in the History of Science and Technology in China*, edited by Li Guohao et al., 659–90. Shanghai: Chinese Classics, 1982.

———. "Telling Famine Stories: The Wanli Emperor and the 'Henan Famine' of 1594." *Études chinoises* 34, no. 2 (2015): 163–202.

———. "Trade and Conflict in the South China Sea: China and Portugal, 1514–1523." In *A Global History of Trade and Conflict since 1500*, edited by Lucia Coppolaro and Francine McKenzie, 20–37. Basingstoke: Palgrave Macmillan, 2013.

———. "Trading Places." *Apollo*, November 2015, 70–74.

———. *The Troubled Empire: China in the Yuan and Ming Dynasties*. Cambridge, MA: Harvard University Press, 2010.

———. *Vermeer's Hat: The Seventeenth Century and the Dawn of the Global World*. New York: Bloomsbury, 2008.

Campbell, Bruce, *The Great Transition: Climate, Disease and Society in the Late-Medieval World*. Cambridge: Cambridge University Press, 2016.

Cartier, Michel. "Les importations de métaux monétaires en Chine: Essai sur la conjoncture chinoise." *Annales* 35, no. 3 (1981): 454–66.

———. "Note sur l'histoire des prix en Chine du XIVe au XVIIe siècle." *Annales* 24, no. 4 (1969): 876–79.

———. *Une réforme locale en Chine au XVIe siècle: Hai Rui à Chun'an, 1558–1562*. Paris: Mouton, 1973.

Chang, Pin-tsun. "The Sea as Arable Fields: A Mercantile Outlook on the Maritime Frontier of Late Ming China." In *The Perception of Maritime Space in Traditional Chinese Sources*, edited by Angela Schottenhammer and Roderich Ptak, 12–267. Wiesbaden: Harrassowitz, 2006.

Chaudhuri, K. N. *The Trading World of Asia and the East India Company, 1660–1760*. Cambridge: Cambridge University Press, 1978.

Chen Gaoyong 陳高傭. *Zhongguo lidai tianzai renhuo biao* 中國歷代天災人禍表 [A list of natural and human disasters in Chinese history]. Shanghai: Jinan daxue, 1939. Shanghai: Shanghai shudian, 1986.

Chen Xuewen 陳學文, ed. *Huzhou fu chengzhen jingji shiliao leizuan* 湖州府城鎮經濟史料類纂 [Collected historical materials on the urban economy of Huzhou prefecture]. Hangzhou, 1989.

Cheng, Hai, Lawrence Edwards, and Gerald Haug. "Comment on 'On Linking Climate to Chinese Dynastic Change: Spatial and Temporal Variations of Monsoonal Rain.'" *Chinese Science* 55, no. 32 (November 2010): 3734–37.

Cheng Minsheng 程民生. *Songdai wujia yanjiu* 宋代物價研究 [Studies in the price history of the Song dynasty]. Beijing: Renmin chubanshe, 2008.

Chow, Kai-wing. *Publishing, Culture, and Power in Early Modern China*. Stanford, CA: Stanford University Press, 2004.

Chuan, Han-sheng, and Richard A. Kraus. *Mid-Ch'ing Rice Markets and Trade: An Essay in Price History*. Cambridge, MA: East Asian Research Center, Harvard University, 1975.

Clunas, Craig. "The Art Market in 17th Century China: The Evidence of the Li Rihua Diary." *History of Art and History of Ideas* 美術史與觀念史, no. 1 (Nanjing: Nanjing shifan daxue chubanshe, 2003): 201–24.

———. *Elegant Debts: The Social Art of Wen Zhengming*. London: Reaktion Books, 2004.

———. *Fruitful Sites: Garden Culture in Ming Dynasty China*. London: Reaktion Books, 1996.

———. *Screen of Kings: Royal Art and Power in Ming China*. Honolulu: University of Hawai'i Press, 2013.

———. *Superfluous Things: Material Culture and Social Status in Early Modern China*. Cambridge: Polity, 1991.

Coatsworth, John H. "Economic History and the History of Prices in Colonial Latin America." In *Essays on the Price History of Eighteenth-Century Latin America*, edited by Lyman Johnson and Enrique Tandeter, 21–33. Albuquerque: University of New Mexico Press, 1990.

Cook, Harold. *Matters of Exchange: Commerce, Medicine, and Science in the Dutch Golden Age*. New Haven, CT: Yale University Press, 2007.

Crosby, Alfred. *The Columbian Exchange: Biological and Cultural Consequences of 1492*. Westport, CT: Greenwood, 1972.

Dai, Lianbin. "The Economics of the Jiaxing Edition of the Buddhist Tripitaka." *T'oung pao* 24, no. 4/5 (2008): 306–59.

Dardess, John. *Four Seasons: A Ming Emperor and His Grand Secretaries in Sixteenth-Century China*. Lanham, MD: Rowman and Littlefield, 2016.

David, Percival, trans. *Chinese Connoisseurship: The Ko Ku Yao Lun, the Essential Criteria of Antiquities*. London: Faber and Faber, 1971.

Davis, Mike. *Late Victorian Holocausts: El Niño Famines and the Making of the Third World*. London: Verso, 2002.

Degroot, Dagomar. *The Frigid Golden Age: Climate Change, the Little Ice Age, and the Dutch Republic, 1560–1720*. Cambridge: Cambridge University Press, 2018.

Deng, Kent. "Miracle or Mirage? Foreign Silver, China's Economy and Globalization from the Sixteenth to the Nineteenth Centuries." *Pacific Economic Review* 13, no. 3 (2008): 320–58.

De Vries, Jan. *The Price of Bread: Regulating the Market in the Dutch Republic.* New York: Cambridge University Press, 2019.

Dudink, Adrian. "Christianity in Late Ming China: Five Studies." PhD diss., University of Leiden, 1995.

Dunstan, Helen. "The Late Ming Epidemics: A Preliminary Survey." *Ch'ing-shih wen-t'i* 3, no. 3 (November 1975): 1–59.

Dyer, Christopher. *Standards of Living in the Later Middle Ages: Social Change in England c. 1200–1520.* Cambridge: Cambridge University Press, 1989.

Dyer, Svetlana Rimsky-Korsakoff. *A Grammatical Analysis of the "Lao Ch'i-ta" with an English Translation of the Chinese Text.* Canberra: Faculty of Asian Studies, Australian National University, 1983.

Ebrey, Patricia. *Chinese Civilization and Society: A Sourcebook.* New York: Free Press, 1981.

Edvinsson, Rodney, and Johan Söderberg. "The Evolution of Swedish Consumer Prices, 1290–2008." In *Exchange Rates, Prices, and Wages, 1277–2008,* edited by Rodney Edvinsson et al., 412–52. Stockholm: Ekerlids Förlag, 2010.

Farmer, Edward. *Zhu Yuanzhang and Early Ming Legislation: The Reordering of Chinese Society following the Era of Mongol Rule.* Leiden: Brill, 1995.

Filipiniana Book Guild. *The Colonization and Conquest of the Philippines by Spain: Some Contemporary Source Documents, 1559–1577.* Manila: Filipiniana Book Guild, 1965.

Finlay, Robert. *The Pilgrim Art: Cultures of Porcelain in World History.* Berkeley: University of California Press, 2010.

Fischer, David Hackett. *The Great Wave: Price Revolutions and the Rhythm of History.* New York: Oxford University Press, 1996.

Flynn, Dennis, and Arturo Giráldez. "Born with a 'Silver Spoon': The Origin of World Trade in 1571." *Journal of World History* 6, no. 2 (Fall 1995): 201–21.

Frank, Andre Gunder. *ReOrient: Global Economy in the Asian Age.* Berkeley: University of California Press, 1998.

Fu Yiling 傅衣凌. "Lun Ming-Qing shidai fengjian tudi maimai qiyue zhong de yinzhu" 論明清時代封建土地買賣契約中的銀主 [On the silver master in feudal land purchase contracts during the Ming-Qing period]. *Dousou* 52 (1983).

Gallagher, Louis, ed. *China in the Sixteenth Century: The Journals of Matthew Ricci, 1583–1610.* New York: Random House, 1953.

Ge Quansheng, Jingyun Zheng, Yanyu Tian, Wenxiang Wu, Xiuqi Fang, and Wei-Chyung Wang. "Coherence of Climatic Reconstruction from Historical Documents in China by Different Studies." *International Journal of Climatology* 28, no. 8 (2008): 1007–24.

Gerritsen, Anne. *The City of Blue and White: Chinese Porcelain and the Early Modern World.* Cambridge: Cambridge University Press, 2020.

Gibson, A.J.S., and T. C. Smout. *Prices, Food and Wages in Scotland, 1550–1780.* Cambridge: Cambridge University Press, 1995.

Gil, Juan. *Los Chinos en Manila (siglos XVI y XVII).* Lisboa: Centro Cientifico e Cultural de Macau, 2011.

Grass, Noa. "Revenue as a Measure for Expenditure: Ming State Finance before the Age of Silver." PhD diss., University of British Columbia, 2015.

Grove, Jean. "The Onset of the Little Ice Age." In *History and Climate: Memories of the Future*, edited by P. D. Jones et al., 153–85. New York: Kluwer, 2001.

Guangzhou shi wenwu guanlichu 廣州市文物管理處 (Cultural Objects Management Office of Guangzhou Municipality). "Guangzhou Dongshan Ming taijian Wei Juan mu qingli jianbao" 廣州東山明太監韋眷墓清理簡報 [Summary report on the grave of Ming eunuch Wei Juan at Dongshan, Guangzhou]. *Kaogu* 1977, no. 4:280–83.

Guo, Yanlong. "Affordable Luxury: The Entanglements of the Metal Mirrors in Han China (202 BCE–220 CE)." PhD diss., University of British Columbia, 2016.

Hamashima Atsutoshi 濱島敦俊. "Minmatsu Kōnan kyōshin no gutaisō" [A concrete image of the gentry of late Ming Jiangnan]. In *Minmatsu Shinsho ki no kenkyū* 明末清初期の研究 [Studies in the period of the late Ming and early Qing], edited by Iwami Hiroshi 岩見宏, 165–83. Kyoto: Kyōto daigaku jimbun kagaku kenkyūjo, 1989.

Hamilton, Earl. "American Treasure and the Rise of Capitalism." *Economica* 27 (1929): 338–57.

———. "Use and Misuse of Price History." *Journal of Economic History* 4, Supplement (December 1944): 47–60.

Harris, Jonathan Gil. *Sick Economies: Drama, Mercantilism, and Disease in Shakespeare's England*. Philadelphia: University of Pennsylvania Press, 2011.

Hegel, Robert. "Niche Marketing for Late Imperial Fiction." In *Printing and Book Culture in Late Imperial China*, edited by Cynthia J. Brokaw and Kai-wing Chow, 236–37. Berkeley: University of California Press, 2005.

Heijdra, Martin (Ma Tailai 馬泰來). "Mingdai wenwu dagu Wu Ting shilüe" 明代文物大賈吳廷事略 [A brief account of Wu Ting, a major dealer in cultural objects in the Ming]. *Gugong xueshu jikan* 23, no. 1 (2005): 397–411.

Ho, Chui-mei. "The Ceramic Trade in Asia, 1602–82." In *Japanese Industrialization and the Asian Economy*, edited by A.J.H. Latham and Heita Kawakatsu, 35–70. London: Routledge, 1994.

Ho, Ping-ti. *The Ladder of Success in Traditional China: Aspects of Social Mobility, 1368–1911*. New York: Columbia University Press, 1962.

Horesh, Niv. "Chinese Money in Global Context: Historic Junctures between 600 BCE and 2012." Stanford Scholarship Online, doi:10.11126/stanford/9780804787192.003.0004. Translated from "The Great Money Divergence: European and Chinese Coinage before the Age of Steam," *Zhongguo wenhua yanjiusuo xuebao* 中國文化研究所學報 [Journal of Chinese studies] 55 (July 2012): 103–37.

Huang Miantang 黃冕堂. *Mingshi guanjian* 明史管見 [Observations on Ming history]. Jinan: Qi Lu shushe, 1985.

Huang, Ray. *1587, a Year of No Significance*. New Haven, CT: Yale University Press, 1981.

———. *Taxation and Governmental Finance in Sixteenth-Century Ming China*. Cambridge: Cambridge University Press, 1974.

Huang Yu 黃煜. *Bixue lu* 碧血錄 [Record of blood]. Zhibuzu zhai congshu ed. Shanghai: Gushu liutongchu, 1921.

Huang Zhangjian 黃彰健. "Ming Hongwu Yongle chao de bangwen junling" 明洪武永樂朝的榜文軍令 [Proclamations and orders the Hongwu and Yongle eras of the Ming]. Reprinted in

Huang Zhangjian, *Ming-Qing shi yanjiu conggao* 明清史研究叢稿 [Research essays in Ming-Qing history], 237–86. Taipei: Shangwu yinshuguan, 1977.

Inoue Susumu 井上進. *Chūgoku shuppan bunka shi: shomotsu sekai to chi no fūkei* 中国出版文化史：書物世界と知の風景 [A cultural history of publishing of China: The book world and the landscape of learning]. Nagoya: Nagoya University Press, 2002.

Jiang, Yonglin. "Defending the Dynastic Order at the Local Level: Central-Local Relations as Seen in a Late-Ming Magistrate's Enforcement of the Law." *Ming Studies* 1 (2000): 16–39.

———, trans. *The Great Ming Code / Da Ming lü*. Seattle: University of Washington Press, 2005.

Jiangxi sheng qinggongye ting taoci yanjiu (Porcelain Research Group of the Jiangxi Provincial Light Industry Bureau), ed. *Jingdezhen taoci shigao* 景德鎮陶瓷史稿 [Draft history of Jingdezhen porcelains]. Beijing: Sanlian shudian, 1959.

Kaplan, Edward, trans. *A Monetary History of China*. 2 vols. Bellingham: Center for East Asian Studies, Western Washington University, 1994.

Kawakatsu Mamoru 川勝守. *Min-Shin Kōnan nōgyō keizai shi kenkyū* 明清江南經濟史研究 [Studies in the history of the agricultural economy of Jiangnan in the Ming and Qing]. Tokyo: Tōkyō daigaku shuppankai, 1992.

Kindleberger, Charles. *Historical Economics: Art or Science?* Berkeley: University of California Press, 1990.

King, Gail, trans. "The Family Letters of Xu Guangqi." *Ming Studies* 21 (Spring 1991): 1–41.

Kishimoto Mio 北本美緒 (*see also* Nakayama Mio). "Minmatsu dendo no shijō ni kansuru ichi kōsatsu" 明末田土の市場關一考察. In *Yamane Yukio kyōju taikyū kinen Mindaishi ronsō* 山根幸夫教授退休紀念明代史論叢 [Essays on Ming history to commemorate the retirement of Yamane Yukio], 2:751–70. Tokyo: Kyūko shoin, 1990.

———. *Shindai Chūgoku no bukka to keizai hendō* 清代中國の物價と經濟變動 [Prices and economic change in Qing China]. Tokyo: Kenbun shuppan, 1997.

Klein, Herbert S., and Stanley J. Engerman. "Methods and Meanings in Price History." In *Essays on the Price History of Eighteenth-Century Latin America*, edited by Lyman L. Johnson and Enrique Tandeter, 9–20. Albuquerque: University of New Mexico Press, 1990.

Kueh, Y. Y. *Agricultural Instability in China, 1931–1991: Weather, Technology, and Institutions*. Oxford: Clarendon, 1995.

Kuo, Jason. "Huizhou Merchants as Art Patrons in the Late Sixteenth and Early Seventeenth Centuries." In *Artists and Patrons: Some Economic and Economic Aspects of Chinese Painting*, edited by Li Chu-tsing, 177–88. Seattle: University of Washington Press, 1989.

Kuroda Akinobu. "What Can Prices Tell Us about 16th–18th Century China?" *Chūgoku shigaku* 中国史学 13 (2003): 101–17.

Lavin, Mary. *Mission to China: Matteo Ricci and the Jesuit Encounter with the East*. London: Faber, 2011.

Lee, Fabio Yu-ching, and José Luis Caño Ostigosa. *Studies on the Map "Ku Chin Hsing Sheng Chih Tu."* Taipei: Research Center for Humanities and Social Sciences, National Tsing Hua University, 2017.

Le Goff, Jacques. *Money and the Middle Ages: An Essay in Historical Anthropology*. Translated by Jean Birrell. Cambridge: Polity, 2012.

Le Roy Ladurie, Emmanuel. "The Birth of Climate History." In *Climate Change and Cultural Transformation in Europe*, edited by Claus Leggewie and Franz Mauelshagen, 197–215. Leiden: Brill, 2018.

————. *Histoire humaine et comparée du climat*. Vol. 1, *Canicules et glaciers (XIIIe–XVIIIe siècle)*. Paris: Fayard, 2004.

Li Defu 李德甫. *Mingdai renkou yu jingji fazhan* 明代人口與經濟發展 [Ming population and economic development]. Beijing: Zhongguo shehui kexue chubanshe, 2008.

Li Guimin 李貴民. "Ming-Qing shiqi landianye yanjiu" 明清時期藍靛業研究]Studies in the indigo industry in the Ming-Qing period]. MA diss., Guoli chenggong daxue, Taipei, 2004.

Li Jiannong 李劍農. "Mingdai de yige guanding wujiabiao yu buhuan zhibi" 明代的一個官定物價表與不換紙幣 [A Ming official price list and inconvertible paper currency]. Reprinted in *Mingdai jingji* 明代經濟 [The Ming economy], vol. 8 of *Mingshi luncong* 明史論叢 [Essays on Ming history], 247–67. Taipei: Xuesheng shuju, 1968.

Li Zichun 李子春. "Mingdai yijian youguan wujia de shiliao" 明代一件有关物价的史料 [A historical document from the Ming dynasty regarding commodity prices]. *Kaogu* 1960, no. 10:50.

Liang Fangzhong 梁方仲, ed. *Zhongguo lidai hukou, tiandi, tianfu tongji* 中國歷代戶口, 田地, 田賦統計 [Historical statistics on Chinese population, land, and taxes]. Shanghai: Renmin chubanshe, 1980.

Liang Jiamian 梁家勉, ed. *Xu Guangqi nianpu* 徐光啟年譜 [Chronological biography of Xu Guangqi]. Shanghai: Shanghai guji chubanshe, 1981.

Libbrecht, Ulrich. *Chinese Mathematics in the Thirteenth Century: The Shu-shu chiu-chang of Ch'in Chiu-shao*. Cambridge, MA: MIT Press, 1973.

Liu Jian et al. "Simulated and Reconstructed Winter Temperatures in the Eastern China during the Last Millennium." *Chinese Science Bulletin* 50, no. 24 (December 2005): 2872–77.

Lu, Tina. "The Politics of Li Yu's *Xianqing ouji*." *Journal of Asian Studies* 81, no. 3 (August 2022): 493–506.

Ma Tailai 馬泰來. *See* Martin Heijdra.

Mann, Michael, et al. "Global Signatures and Dynamical Origins of the Little Ice Age and Medieval Climate Anomaly." *Science* 326 (November 2009): 1256–60.

Marks, Robert. *China: Its Environment and History*. Lanham, MD: Rowman and Littlefield, 2012.

————. "'It Never Used to Snow': Climate Variability and Harvest Yields in Late-Imperial South China, 1650–1850." In *Sediments of Time: Environment and Society in Chinese History*, edited by Mark Elvin and Liu Ts'ui-jung, 435–44. Cambridge: Cambridge University Press, 1998.

————. "Rice Prices, Food Supply, and Market Structures in Eighteenth-Century South China." *Late Imperial China* 12, no. 2 (December 1991): 64–116.

Martzloff, Jean-Claude. *A History of Chinese Mathematics*. Translation of *Histoire des mathématiques chinoises* (1987). New York: Springer, 2006.

Morse, Hosea Ballou. *The Chronicles of the East India Company, Trading to China 1635–1834*. Vol. 1. Oxford: Clarendon, 1926.

Muldrew, Craig. *The Economy of Obligation: The Culture of Credit and Social Relations in Early Modern England*. London: Macmillan, 1998.

————. *Food, Energy and the Creation of Industriousness: Work and Material Culture in Agrarian England, 1550–1780*. Cambridge: Cambridge University Press, 2011.

Munro, John. "Money, Prices, Wages, and 'Profit Inflation' in Spain, the Southern Netherlands, and England during the Price Revolution Era, ca. 1520–ca. 1650." *História e Economia* 4, no. 1 (2008): 14–71.

———. Review of David Hackett Fischer, *The Great Wave: Price Revolutions and the Rhythm of History*. *EH.Net Review*, 24 February 1999, ehreview@eh.net, accessed 10 June 2022.

Nakayama Mio 中山美緒 (*see also* Kishimoto Mio). "On the Fluctuation of the Price of Rice in the Chiang-nan Region during the First Half of the Ch'ing Period (1644–1795)." *Memoirs of the Research Department of the Toyo Bunko* 37 (1979): 55–90.

———. "Shindai zenki Kōnan no bukka dōkō" 清代前期江南の物價動向 [On the rise and fall of commodity prices in the Jiangnan region during the first half of the Qing period]. *Toyoshi kenkyu* 37, no. 4 (1979): 77–106. Reprinted in Kishimoto Mio, *Shindai Chūgoku no bukka to keizai hendō* 清代中國の物價と經濟變動 [Prices and economic change in Qing China], 99–135. Tokyo: Kenbun shuppan, 1997.

Niida Noboru. *Chūgoku hōseishi kenkyū: Dorei nōdo hō, kazoku sonraku hō* 中國法制史研究: 奴隷農奴法, 家族村落法 [Studies in the history of Chinese law: Slave and surf laws, family and village law]. Tokyo: Tōkyō daigaku shuppansha, 1981.

Oertling, Sewall. *Painting and Calligraphy in the "Wu-tsa-tsu": Conservative Aesthetics in Seventeenth-Century China*. Ann Arbor: Center for Chinese Studies, University of Michigan, 1997.

Paethe, Cathleen, and Dagmar Schäfer. "Books for Sustenance and Life: Bibliophile Practices and Skills in the Late Ming and Qi Chenghan's Library Dasheng Tang." In *Transforming Book Culture in China, 1600–2016* (Kodex 6), edited by Daria Berg and Giorgio Strafella, 19–48. Wiesbaden: Harrassowitz Verlag, 2016.

Parker, Geoffrey. *Global Crisis: War, Climate Change and Catastrophe in the Seventeenth Century*. New Haven, CT: Yale University Press, 2013.

———. "History and Climate: The Crisis of the 1590s Reconsidered." In *Climate Change and Cultural Transformation in Europe*, edited by Claus Leggewie and Franz Mauelshagen, 119–55. Leiden: Brill, 2018.

Parsons, James. *Peasant Rebellions of the Late Ming*. Tucson: University of Arizona Press, 1970.

Peng Xinwei. *Zhongguo huobi shi* 中國貨幣史 [A monetary history of China]. Shanghai: Qunlian chubanshe, 1954.

Pomeranz, Kenneth. *The Great Divergence: China, Europe, and the Making of the Modern World Economy*. Princeton, NJ: Princeton University Press, 2000.

Prange, Sebastian. "'Measuring by the Bushel': Reweighing the Indian Ocean Pepper Trade." *Historical Research* 84, no. 224 (May 2011): 212–35.

Qin Peiheng 秦佩珩. "Mingdai mijia kao" 明代米價考 [Notes on grain prices in the Ming dynasty]. In Qin Peiheng, *Ming-Qing shehui jingji shi lungao* 明清社會經濟史論稿 [Draft essays on socioeconomic history of the Ming and Qing], 199–210. Zhengzhou: Zhongzhou guji chubanshe, 1984.

Quan Hansheng 全漢昇. *Ming Qing jingji shi yanjiu* 明清經濟史研究 [Studies in the economic history of the Ming and Qing]. Taipei: Lianjing, 1987.

———. "Song Ming jian baiyin goumaili de biandong ji qi yuanyin" 宋明間白銀購買力的變動及其原因 [Changes in the purchasing power of silver from Song to Ming and their causes]. *Xinya xuebao* 8, no. 1 (1967): 157–86.

Quanzhou shi wenwu guanli weiyuanhui 泉州市文物管理委員會 (Quanzhou Cultural Artifacts Management Committee) and Quanzhou shi haiwai jiaotongshi bowuguan 泉州市海外交通史博物馆 (Quanzhou Museum of Overseas Communications). "Fujian Quanzhou diqu chutu de wupi waiguo yinbi" 福建泉州地區出土的五批外國銀币 [Five batches of foreign silver coins excavated in the Quanzhou region, Fujian]. *Kaogu* 1975, no. 6:373–80.

Reddy, William. *Money and Liberty in Modern Europe: A Critique of Historical Understanding.* Cambridge: Cambridge University Press, 1987.

Satō Taketoshi 佐藤武敏. *Chūgoku saigaishi nenpyō* 中国災害史年表 [Annual chronology of historical disasters in China]. Tokyo: Kokusho kankōkai, 1993.

Rusk, Bruce. "Value and Validity: Seeing through Silver in Late Imperial China." In *Powerful Arguments: Standards of Validity in Late Imperial China*, edited by Martin Hofmann et al., 471–500. Leiden: Brill, 2020.

Schäfer, Dagmar. *The Crafting of the Ten Thousand Things: Knowledge and Technology in Seventeenth-Century China.* Chicago: University of Chicago Press, 2011.

Schäfer, Dagmar, and Dieter Kuhn. *Weaving and Economic Pattern in Ming Times (1368–1644): The Production of Silk Weaves in the State-Owned Silk Workshops.* Heidelberg: Edition Forum, 2002.

Shan Kunqing. "Copper Cash in Chinese Short Stories Compiled by Feng Menglong (1574–1646)." In *Money in Asia (1200–1900): Small Currencies in Social and Political Contexts*, edited by Jane Kate Leonard and Ulrich Theobald, 224–46. Leiden: Brill, 2015.

Siebert, Lee, Tom Simkin, and Paul Kimberley. *Volcanoes of the World.* 3rd ed. Berkeley: University of California Press, 2011.

Song Zhenghai 宋正海. *Zhongguo gudai ziran zaiyi xiangguanxing nianbiao zonghui* 中國古代自然災異相關性年表總匯 [Combined chronology of natural disasters in ancient China]. Hefei: Anhui jiaoyu chubanshe, 2002.

Struve, Lynn. *Voices from the Ming-Qing Cataclysm: China in Tigers' Jaws.* New Haven, CT: Yale University Press, 1993.

Su Gengsheng 蘇更生. "Mingchu de shangzheng yu shangshui" 明初的商政與商稅 [Merchant policy and commercial taxation in the early Ming]. *Mingshi yanjiu luncong* 明史研究論叢 [Research essays on Ming history] 1985, no. 2:427–48.

Torres, José Antonio Martinez. "'There Is But One World': Globalization and Connections in the Overseas Territories of the Spanish Hapsburgs (1581–1640)." *Culture and History Digital Journal* 3, no. 1 (June 2014). https://brasilhis.usal.es/en/node/7660.

Tsien, Tsuen-Hsuin. *Science and Civilisation in China.* Vol. 5:1, *Paper and Printing.* Cambridge: Cambridge University Press, 1985.

Volker, T. *Porcelain and the Dutch East India Company, as Recorded in the Dagh-Registers of Batavia Castle, Those of Hirado and Deshima and Other Contemporary Papers, 1602–1682.* Leiden: Brill, 1954.

von Glahn, Richard. "The Changing Significance of Latin American Silver in the Chinese Economy, 16th–19th Centuries." *Journal of Iberian and Latin American Economic History* 38, no. 3 (December 2020): 553–85.

———. *Fountain of Fortune: Money and Monetary Policy in China, 1000–1700.* Berkeley: University of California Press, 1996.

————. "Money Use in China and Changing Patterns of Global Trade in Monetary Metals, 1500–1800." In *Global Connections and Monetary History, 1470–1800*, edited by Dennis Flynn, Arturo Giráldez, and Richard von Glahn, 187–205. Burlington: Ashgate, 2003.

Wakeman, Frederic. *The Great Enterprise: The Manchu Reconstruction of Imperial Order in Seventeenth-Century China*. Berkeley: University of California Press, 1985.

Wang Guangyao 王光堯. *Mingdai gongting taoci shi* 明代宮廷陶瓷史 [History of Ming court porcelain]. Beijing: Zijincheng chubanshe, 2010.

Wang Jiafa 王家範. "Ming-Qing Jiangnan xiaofei jingji tance" 明清江南消費經濟探測 [Exploration of the Jiangnan consumer economy in the Ming and Qing]. *Huadong shifan daxue xuebao* 1988, no. 2:157–67.

Wang Teh-yi 王德毅, ed. *Zhonghua minguo Taiwan diqu gongcang fangzh mului* 中華民國臺灣地區公藏方志目錄 [Union catalogue of Chinese gazetteers in public collections in Taiwan, ROC]. Taipei: Resource and Information Center for Chinese Studies, 1985.

Wang, Yeh-chien. "The Secular Trend of Prices during the Ch'ing Period (1644–1911)." *Zhongguo wenhua yanjiusuo xuebao* [Journal of the Institute of Chinese Culture] 5, no. 2 (1972): 347–71.

————. "Secular Trends of Rice Prices in the Yangzi Delta, 1638–1935." In *Chinese History in Economic Perspective*, edited by Thomas Rawsi and Lillian Li, 35–68. Berkeley: University of California Press, 1992.

Ward, Peter. "Stature, Migration and Human Welfare in South China, 1850–1930." *Economics and Human Biology* 11, no. 4 (December 2013): 488–501.

Wilkinson, Endymion. *Studies in Chinese Price History*. New York: Garland, 1980.

Will, Pierre-Étienne. "Discussions about the Market-Place and the Market Principle in Eighteenth-Century Guangdong." *Zhongguo haiyang fazhanshi lunwen ji* 中國海洋發展史論文集 7 (1999): 323–89.

————. *Handbooks and Anthologies for Officials in Imperial China: A Descriptive and Critical Bibliography*. Leiden: Brill, 2020.

Will, Pierre-Étienne, and R. Bin Wong, eds. *Nourish the People: The State Civilian Granary System in China, 1650–1850*. Ann Arbor: Center for Chinese Studies, University of Michigan, 1991.

Wilson, Rob, et al. "Last Millennium Northern Hemisphere Summer Temperatures from Tree Rings: Part I. The Long Term Context." *Quaternary Science Reviews*, 9 January 2016, http://dx .doi.org/10.1016/j.quascirev.2015.12.005.

Wong, R. Bin. *China Transformed: Historical Change and the Limits of European Experience*. Ithaca, NY: Cornell University Press, 1997.

Wu Chengluo 吳承洛. *Zhongguo duliangheng shi* 中國度量衡史 [A history of Chinese weights and measures]. 1937. Shanghai: Shangwu yinshuguan, 1957.

Wu Renshu 巫仁恕. *Pinwei shehua: Wan Ming de xiaofei shehui yu shidafu* 品味奢華: 晚明的消費社會與士大夫 [Taste and extravagance: Late Ming consumer society and the gentry]. Taipei: Zhongyang yanjiuyuan lianjing chuban gongsi, 2007.

————. *Youyou fangxiang: Ming Qing Jiangnan chengshi de xiuxian xiaofei yu kongjian bianqian* 優游坊廂: 明清江南城市的休閒消費與空間變遷 [Urban pleasures: leisure consumption and spatial transformation in Jiangan cities during the Ming-Qing period]. Taipei: Zhongyang yanjiuyuan jindaishi yanjiusuo, 2013.

Xiao, Lingbo, Xiuqi Fang, Jingyun Zheng, and Wanyi Zhao. "Famine, Migration and War: Comparison of Climate Change Impacts and Social Responses in North China in the Late Ming and Late Qing Dynasties." *Holocene* 25, no. 6 (2015): 900–910.

Xu Hong 徐泓. "Jieshao jize Wanli sishisan, si nian Shandong jihuang daozhi renxiangshi de shiliao" 介紹幾則萬曆四十三, 四年山東饑荒導致人相食的史料 [Introducing materials on several cases of cannibalism in the Shandong famine of 1615–16]. *Mingdai yanjiu tongxun* 6 (2003): 143–49.

——. "Mingmo shehui fengqi de bianqian—yi Jiang, Zhe diqu wei li" 明末社會風氣的變遷—以江浙地區為例 [Changes in social customs in the late Ming, taking the Jiangsu-Zhejiang region as an example]. *Dongya wenhua* 24 (1986): 83–110.

Xue Longchun 薛龍春. *Wang Duo nianpu changbian* 王鐸年譜長編 [Extended chronological biography of Wang Duo], 3 vols. Beijing: Zhonghua shuju, 2019.

Yang Lien-sheng. *Money and Credit in China: A Short History*. Cambridge, MA: Harvard University Press, 1952.

Ye Kangning 葉康寧. *Fengya zhi hao: Mingdai Jia-Wan nianjian de shuhua xiaofei* 風雅之好: 明代嘉萬年間的書畫消費 [What style was favored: The consumption of painting and calligraphy in the years of the Jiajing to Wanli eras of the Ming dynasty]. Beijing: Shangwu yinshuguan, 2017.

Yim, Shui-yuen. "Famine Relief Statistics as a Guide to the Population of Sixteenth-Century China: A Case Study of Honan Province." *Ch'ing-shih wen-t'i* 3, no. 9 (1978): 1–30.

Zhang Anqi. "Ming gaoben 'Yuhua tang riji' zhong de jingjishi ziliao yanjiu" 明末稿本玉華堂日記的經濟史資料研究 [Study of materials for economic history in the Ming edition of "Diary from Jade Flower Hall"]. In *Mingshi yanjiu luncong*, 5:268–311. Nanjing: Jiangsu guji chubanshe, 1991.

Zhang Jiacheng. *The Reconstruction of Climate in China for Historical Times*. Beijing: Science Press, 1988.

Zhang Jiacheng and Thomas Crowley. "Historical Climate Records in China and Reconstruction of Past Climates." *Journal of Climate* 2 (August 1989): 833–49.

Zheng, Jingyun, Lingbo Xiao, Xiuqi Fang, Zhixin Hao, Quansheng Ge, and Beibei Li. "How Climate Change Impacted the Collapse of the Ming Dynasty." *Climatic Change* 127, no. 2 (2014): 169–82.

Zhongyang qixiang ju qixiang kexue yanjiu yuan 中央氣象局氣象科學研究院, ed. *Zhongguo jin wubai nian hanlao fenbu tuji* 中國近五百年旱澇分布圖 [Maps of the distribution of droughts and pluvials in China over the last five hundred years]. Beijing: Ditu chubanshe, 1981.

INDEX

Page numbers in *italics* denote figures or tables.

A NOTE ON THE TYPE

This book has been composed in Arno, an Old-style serif typeface in the classic Venetian tradition, designed by Robert Slimbach at Adobe.